Beeconomy

Beeconomy

*What Women and Bees
Can Teach Us about Local Trade
and the Global Market*

Tammy Horn

THE UNIVERSITY PRESS OF KENTUCKY

Scholarly publisher for the Commonwealth,
serving Bellarmine University, Berea College, Centre
College of Kentucky, Eastern Kentucky University,
The Filson Historical Society, Georgetown College,
Kentucky Historical Society, Kentucky State University,
Morehead State University, Murray State University,
Northern Kentucky University, Transylvania University,
University of Kentucky, University of Louisville,
and Western Kentucky University.
All rights reserved.

Editorial and Sales Offices: The University Press of Kentucky
663 South Limestone Street, Lexington, Kentucky 40508-4008
www.kentuckypress.com

16 15 14 13 12   5 4 3 2 1

Library of Congress Cataloging-in-Publication Data

Horn, Tammy, 1968–
  Beeconomy : what women and bees can teach us about local trade and the
global market / Tammy Horn.
      p. cm.
  Includes bibliographical references and index.
  ISBN 978-0-8131-3435-2 (hardcover : alk. paper)—
  ISBN 978-0-8131-3436-9 (ebook)
  1. Women beekeepers. 2. Women in agriculture. 3. Bee culture. I. Title.
II. Title: What women and bees can teach us about local trade and the global
market.
  SF524.H67   2011
  638′.1092—dc23
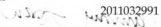   2011032991

This book is printed on acid-free paper meeting
the requirements of the American National Standard
for Permanence in Paper for Printed Library Materials.

Manufactured in the United States of America.

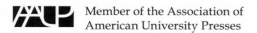 Member of the Association of
American University Presses

For Charlene Hacker and Earl Horn,
who taught me the importance of the past

For Elaine and Edwin Holcombe,
who support me in the present

For Dr. Alice Jones and Dr. Richard Scott,
who envision with me a brilliant future

I've been told by literary experts that the refined reader who needs plot, pace, action and a body on the hearthrug in the first chapter will recoil from the wearisome and comical statement of our drudgery.

Well by God, I say, rolling up my sleeves, I'm going to write the book and let whoever pleases read it or not just as they prefer. I'm writing it to clear my mind, cutting a track through a forest of bewilderment and lies and laughter. It's my road and you don't need to bring your truck in unless you want a load of honey. I may get bogged, I may have to jack up, and swear at the tough going, doing everything the hard way as usual, but when I do get through, it is to somewhere where the great eucalypts are in flower, and the silver-leaved ironbark is promising, where the bees are in good heart, and the creek runs clear and there is firewood and friendship.

—Kylie Tennant

Contents

Illustrations

Acknowledgments

Ibsen once wrote that he could not afford friendship. I have never known that kind of poverty.

"Friendship" is too simple a term for the debt I owe to many people. Among international circles, numerous people such as Penny Walker, Keith and Trish Chisnall, Richard and Jane Jones, Nicola Bradbear, Kunal Sharma, Stan Glowacki, Gro Amdam, Jeremy Burbidge, Karl Showler, David Clemmons, Elise Sabatini, and Liann McGregor illuminated corners of their homelands in ways that left indelible memories of their kindness.

Key colleagues in the academic community are as follows: James Kenkel, Kevin Jones, Frank Davis, and Cary Hazelwood, who provide bibliographic assistance; Hanna Watts, Robert Browning, and Nan Campbell, who have proved a good-natured field crew; Krista Whitaker and Tiffany Hamblin, who apply their logistic skills to grants with grace and humor; Jennifer Spock, Joyce Harrison, Karrie Atkins, Reagan Weaver, John Perry, Dobree Adams, Jonathan Greene, Bob Shadur, Marc Whitt, Jerry Wallace, Wayne Waddell, Richard Tussey, Tom Britz, Regina Fultz, Ian Mooers, Cassie Williams, Gerald Pogatschnick, and Eastern Kentucky University president Doug Whitlock, who provide a solid team. The University Press of Kentucky continues to provide invaluable guidance, most notably Steve Wrinn, Allison Webster, Mack McCormick, Ashley Runyon, Robert Brandon, Lin Wirkus, and Teresa Collins.

Research for this book was gleaned in the following libraries: Cape Town National Library, Cornell University's Mann Library (the Hive and Honeybee Online Collection), Eastern Kentucky University Crabbe Library, International Bee Research Association, Pennsylvania State University, University of Cape Town, University of Connecticut, University of Kentucky William T. Young and Medical Libraries, University of Wisconsin–Madison Library, Victoria National Library (Australia), and Washington-on-the-Brazos State Park Library.

In an era when editors seem to come and go, Kim Flottum and Joe Graham have continued to provide steady leadership at the helms of *Bee Culture* and *American Bee Journal,* respectively. Claire Waring of *Bee Craft* and John Phipps of *Beekeepers Quarterly* maintain the British posts. The Dadant and Walter T. Kelley Companies have shown great generosity in supplies and information. Kathy Summers and Dawn Feagan complete the much-needed, oft-invisible editorial network.

Philanthropic foundations and the coal industry have provided support for my transition from professor to pollination management. In 2006 the Berea College National Endowment for the Humanities Chair of Appalachian Studies enabled me to research aspects of the queen-rearing production industry in Hawaii and South Africa. In 2007 the Eastern Kentucky Environmental Research Institute, Eastern Kentucky Foundation, the Foundation for the Preservation of Honey Bees, and the Kentucky Foundation for Women provided grants funding extension and outreach. The Steele-Reese Foundation was no less important in its contribution to financial continuity from 2009 to 2011. It has been an honor to represent these institutions when compiling this book.

Other people I work with have been an unexpected element of this book's completion and success. Paul Rothman, Don Graves, Don Gibson, Brian Patton, John Tate, Greg Feltner, Robert Ray, David and Susie Duff, Bob Zik, Stacy Billtner, Scott Kreutzer, and Jonathan Crowie have collaborated with me to define a future in which pollinator habitat becomes a standard best-management practice in surface-mine reclamation.

Acknowledgments

The beekeepers keep me: Tom Webster, Gerald Hayes, Phil Craft, the Bluegrass Beekeepers' Association, the Letcher County Beekeepers' Association, Kent and Valerie Williams, Toni Downs, Lucy Breathitt, Mark and Mary Lee, Annie and Richard Broyles, Donna Jo Horn, Stephanie Tarwater, Staci Adair, Ed Levi, Bill Troupe, Kathy Currier, Allen and Ruby Meyers, Teddy and Judy Jones, Richard and Rita Underhill, Kim Lehman, Clarence Collison, Maryann and Jim Frazier, Diana Cox-Foster, Jennifer Berry, Sean Clark, Sue Cobey, Lucy Breathitt, Diana Sammataro, Patty Pulliam and Serge Boyce, Grace Pundyk, Martin Braunstein and Sonia Verettoni, Gro Amdam, Al Avitabile, Marina Marchese, Norm and Andrew Coté, Craig and Charlene Cella, Abigail Keam, Troy Fore, John Miller, Jay Miller, Amanda Miller, the Olivarez family, Marla Spivak, M. E. A. McNeil, and Greg Hunt challenge me to define new insights. In Hawaii, Joanne Murray and the Ocean View Congregational Church (Papa Kahu, Mama Kahu, Bill and Jim Glover) exemplified the definition of *ohana*.

As always, Gabriel Napier, Jamie Horn, Brian Napier, Lyn Hacker, and E. J. Hacker, as well as Perry Amos, Joyce Henderson, Amy and Sara Noland, Ashley Gibson, Emily Saderholm, Tamara Carmichael Farnsworth, and their loving families provide a solid scaffold.

Sometimes, memories are merely ghosts who no longer frighten or sadden. Because of the powerlessness I feel about her last days, I'd like to acknowledge my grandmother Bess Horn, who kept bees beside the Poor Fork of the Cumberland River. In a country known for its truncated ridges and outcrops, she was soft in the best possible use of that word. How and from whom she learned to keep bees, no one knows.

There had to be a pragmatic quality to her decision, a desire to pollinate her gardens or to circumvent the rationing during World War II. Or it could be she had come to prefer the unconventional companionship, as I would grow to do, sitting in the warm, dark silence, infused in the polleny concoction of wax, honey, and larvae.

This part of my grandmother's past has haunted me since I began to keep bees. The desire to have even a ghostly glimpse of her in this book's photographs has been eased by the book's tales—at times warmly told, at times a struggle for syllables. The effort to represent the women and their stories in the following pages has balanced the reality of my grandmother's last days in an unfamiliar hospital room, broken from trying to get out of an unfamiliar bed. Her death was a struggle.

Because this narrative effort is only half complete without a reader, this acknowledgment is to the reader too. In opening this book, you let my grandmother live as someone other than that anonymous, isolated woman lying helpless in a hospital bed. She lights her smoker, she hums in tune with the bees below . . . these possibilities save me.

Introduction

Piping Up

Women beekeepers have been shortchanged in the beekeeping history books when they actually have made many significant contributions through the years.
—Joe Graham, editor of *American Bee Journal*, 1979

I have been asked, why write a book about women and bees? The subtext of the question is that we surely do not *need* a book about women beekeepers. Nor do I offer any new beekeeping secrets. I am certainly not the best writer on this topic, and neither is my gender considered adequate qualification.

But, with the United States losing one in every three hives of honey bees and Central Europe losing one in four, more women should consider keeping bees.[1] If we have more beekeepers, regardless of gender, perhaps the immediate crisis of bee loss will be addressed and our agricultural sectors will have appropriate pollination to feed the world's citizens.

Colony losses aside, women have much at stake when bee losses are as high as 30 percent. They often have the most direct access to food eaten by family members. They generally live longer than their male counterparts. And women continue to be paid inequitably regardless of location, education, and religion. Some women may benefit from this book because it could prompt them to consider new ways to supplement incomes, improve family nutrition, and enjoy the intangible benefits of beekeeping as an activity.

Whether they are nursing infants, preparing meals for children, or keeping track of medications for elderly parents, women tend to have more links with nutrition than their male counterparts, even though more men in Western societies are participating in domestic responsibilities than ever before. Centuries ago, English writer Mary Wollstonecraft wrote that society should prioritize women's education precisely because mothers are the link between children and nutrition. Healthy mothers will beget healthy children, she surmised. Wollstonecraft intoned to anyone who would listen: "It is vain to expect the present race of weak mothers . . . to take that reasonable care of a child's body, which is necessary to lay the foundation of a good constitution. . . . The weakness of the mother will be visited on the children."[2]

Wollstonecraft first articulated this argument in the 1790s, but the problem of children not having access to affordable fruits and vegetables has had long-term, systemic consequences through the centuries. Even now, the global repercussions of unhealthy diets for children are obesity, childhood diabetes, and dental problems. It is not just the multiple health risks that can be problematic if children do not develop good eating habits early in life; children will also lack the quality of life that comes from eating good food. Children have a right to enjoy biting into blackberries that splatter on their tongues or slurping the sweetness of a watermelon that has been properly pollinated.

I do not think women are deliberately ignorant about the links between pollination and nutrition. Socialization and discrimination have shaped perceptions about women's opportunities in agriculture generally, and specifically in apiculture. Beekeeping can be labor intensive. In honey-hunting cultures such as those in Africa and India, bee activities can be hazardous. Within the past two hundred years, socialization patterns in Western civilizations have reinforced women's choices of steadier, service-oriented positions, such as teachers, nurses, clerks, and retail workers. Agriculture is no longer a reliable step to a middle-class income in some industrial civilizations. Bee-keeping is not approved "women's work" in some theocratic countries, nor is it even, in some societies, a socially acceptable activity.

Discrimination, which can take overt or subtle forms in the apiculture world, has also been a factor, although less so than in other agricultural sectors. Before the twentieth century, restricted admissions to schools and universities discouraged women from the knowledge-based economies associated with apiculture science and extension.

Affirmative action laws have eased some overt discrimination, but generally speaking, salary discrimination continues to affect women, regardless of profession. In 2010, white US women still made seventy-seven cents for every dollar earned by their male counterparts. Black women earned sixty-eight cents for every dollar. According to Merlene Davis, a US-based writer, "The Equal Pay Act has been on the books since 1963, when women were earning 59 cents for every dollar earned by men. So, in the ensuing 47 years, that law has increased our pay only 18 comparative cents. . . . That's less than half a penny a year."[3]

Major issues complicate discrimination in the bee world, including the lack of good-quality, affordable day care, responsible immigration reform, extended social networks, and reliable transportation systems. Hispanic women typically earn less than their white counterparts on the queen-catching crews or in grafting rooms; they are also more likely to have more children at younger ages and thus to have to take time off for child care.

In general, in most careers, not just apiculture, women are more likely than men to be called when their children are sick, to have to take time off work to care for family, and—even with comparable educations—to receive less pay for equal responsibilities. Throughout my own unconventional career, I have resolved these professional discrepancies by understanding how honey bees were used to define femininity thousands of years ago. In this book, women assert a wide range of social roles, encompassing honey gatherers/producers, goddesses, novitiates, editors, authors, extension agents, wax chandlers, scientists, swarm catchers, inspectors, and queen producers. In addition to being beekeepers, women have added to family finances and comfort through value-added industries—cosmetics, candles, or culinary goods, for example. In many contemporary beekeeping families, women run the business side of the operation, performing tasks such as preparing taxes, answering phone calls, shipping orders, and completing paperwork. Women have adapted to so many roles in beekeeping that I have learned to be flexible and creative in approaching my own career.

This book fills some "zero images," a phrase used by scholar Carolyn Gerald in the 1930s to describe the lack of good images for African American children's books.[4] As recently as 2003, Karl Showler, a writer for *Bee Craft* magazine in England, concluded a series of articles about bee books written by women beekeepers, declaring: "We would all benefit from a well-produced modern book about practical beekeeping by a woman for women. . . . Women should be pictured working their bees."[5] If we see women working their hives, it is less likely that we will transfer fallacies from the hive to the home.

The process of reading stories and pictures together—what children's literature scholar Lawrence Sipe calls "oscillation"—challenges assumptions simultaneously and also sequentially. "Just as written language is not purely linear," Sipe writes, "paints and the visual arts are not purely spatial either. . . . It's an intriguing idea that the interrelationship of words and pictures mirrors the thought process itself."[6] Applying this logic to apiculture, we can say most readers immediately associate bees with older white men. Men's achievements in apiculture are well documented and photographed. Not so for women. A chronicle of women's achievements with honey bees is long overdue.

This book also clarifies how women may be socialized by the language regarding honey bees, especially when it comes to queen

Joanne Murray, lighting a smoker on the Big Island, Hawaii. Courtesy of Perry Amos.

bees. If a woman or teenage girl is smart or ambitious, she is often called a "queen bee" without a basic understanding of that role in the hive. Calling a woman a queen bee is not always complimentary. Even in the best of contexts, the label suggests that the woman is powerful, a potential threat, and self-absorbed.

By extension, the term "worker bee" often refers to women who are supposed to labor behind the scenes, underpaid and content to sacrifice for the good of the whole. The label tends to overlook specialization within the hive, and it almost always becomes an expectation for women who are supposed to remain anonymous in their service.

When people adopt these analogies based on female bees, they inadvertently falsify or simplify the relationships between themselves and those within the hive. We use bee-related analogies because they simplify *human* anxieties about women, even if the women happen to be our friends, mothers, sisters, colleagues, and neighbors.

The queen bee (at center) is longer than the workers or drones. Each hive has only one queen, and her pheromones are distributed throughout the hive by her attendants. The drone (to the left) can be distinguished by his larger eyes and more rotund body. Courtesy of *Bee Culture Magazine*.

Leading a shift in language terms, contemporary beekeeper Ann Harmen says she prefers to call a queen bee a "provider," not a ruler. Her distinction is an important clarification to the hive structure. In a hive, the queen bee is a living organism in a complex, ever-shifting arrangement of roles and chemicals that adjusts to a variety of environments. She provides pheromones—chemicals that respond to and control environmental conditions. These pheromones assist the workers in the colony and help the male bees, or "drones," to find her when they mate on the wing. She will mate, on average, with about eight to thirty drones for about two to three weeks, provided the climate is temperate. A queen also provides eggs, laying as many as fifteen hundred to two thousand a day.

Although a queen and worker bees provide much of a hive's needs, a hive still requires drones to mate with the queen and provide

Honey Bee Basics

A queen is only as good as the drones with which she will mate. As maligned as drones are in human society, they are necessary in the bee world to maintain hive health and diversity. Queens generally mate with eight to thirty drones in a region, holding the semen in a storage container called a spermatheca. After a queen returns to the hive, she will begin laying eggs in cells the workers have prepared for her. Measuring the size of the cell during her inspection, the queen determines whether she is in a worker cell. If she is, she releases a small amount of spermathecal contents as she lays an egg in the center; it will become a female worker bee. If the cell is larger, the queen will not release any spermathecal content, and the egg will become a male drone.

As with any living organism, diet is crucial to development. If eggs are placed in specially sized cells, called queen cups, workers will feed those queen eggs copious amounts of royal jelly immediately—in effect, developing queen larvae. Worker bees create queen cups naturally on the bottom of frames, but queen cells can also form on foundation walls.

Eggs in worker bee cells are fertilized (or diploid), meaning that the resulting females will be influenced by the genetic tendencies of both drones and the queen. Compared to the queens, worker bee larvae have lighter feedings of royal jelly and differing concentrations of protein and carbohydrates during the pupae stage. Once emerged from their cells, workers specialize in a variety of tasks as they age, but they generally live no longer than six months. Winter bees live longer than summer bees. A queen bee may live for five years, but some queens live only six months to one year.

When the queen no longer has sperm, has been rendered sterile by pesticides, or has never mated at all, she will lay eggs with no spermathecal content so only drones will result. Drones are fed by worker bees when environmental conditions allow for abundance of resources such as pollen and nectar. Drones exist to mate with queens from other colonies. When worker bees detect the lack of spermathecal content in the worker cells, they begin the process of replacing the queen. (See Connor, *Bee Sex Essentials*.)

genetic diversity. So, true to the economy of the hive, unfertilized eggs, becoming drones, are not wasted. As scientist Jennifer Berry explains, drones are nothing but half a queen.[7] Drones have a shorter lifespan than workers, existing only a few months. Still, studies have shown that the drone pheromone is necessary to maintain the morale of the worker bees throughout the summer. Because they are pushed out of the hive at the end of summer, drones have provided a convenient analogy for human societies to apply to people when economic times get tough, such as when overpopulation becomes an uncomfortable factor.

Although it would be convenient to snap our fingers and make them go away, bee analogies fill a cultural and spiritual vacuum and perhaps shape our expectations regarding women. So desperate was William Wordsworth to retain any belief system that he once begged in a sonnet to be "a Pagan suckled in a creed outworn / So might I, standing on this pleasant lea / Have glimpses that would make me less forlorn."[8] Do false analogies comparing women to queen bees or worker bees *really* do irreparable harm? The easy answer is no. The world is big enough to handle these "outworn creeds," and it is feasible to breathe new life into them when people need something to make them "less forlorn."

But if the analogies create more hierarchies between people, we are remiss in using them. The division between genders is often one created by people, after all, not by divine forces. Author Lynn Margulis in *The Symbiotic Planet* suggests that "language can confuse and deceive." She laments that "antiquated terms . . . remain in use despite their penchant to propagate biological malaise and ignorance," arguing that "these insults to the living benefit people whose budgets, class notes, and social organization depend on their continuity."[9]

Similarly, when women are labeled as queen bees or worker bees, the labels can justify the status quo, whether it be when universities pay women faculty lower salaries than their male colleagues, when school counselors steer females toward a particular profession or a less challenging college, or when women refuse to confront difficult topics because they are afraid they will be labeled aggressive in business or domestic environments. Clearly, women who are not aware of larger socialization patterns may repeat or identify with false analogies. Socialization is not a unilateral process driven by men.

Nor are the patterns affecting women and bees unique to the United States. I have arranged this book geographically to suggest that these socialization patterns are similar around the globe. Such organization loosely replicates the migration of honey bees via human- or swarm-assisted movement. Since Africa is where honey bees first evolved, the book begins with that continent. The second chapter focuses on India's honey-hunting cultures and recent economic development for women.

The following chapter starts with Eastern Asia and concludes with Western Asia. The oldest bee goddesses originated in Anatolia, but until recently, many of the relationships between women and bees had been subsumed by other cultures. However, contemporary political and military events have created links with ancient agricultural ideas. The chapter begins with the Philippines and Malaysia, then explores Russia and its relationships with central Asian countries such as Afghanistan and Iraq, concluding with Turkey.

The chapter on Europe is divided into sections on the south and north. The section on southern Europe explores the Mediterranean region's leadership in bee-related religion and politics during prehistoric times, and the section on northern Europe explores how industrialization provided not only a platform for women to enter the economic system but also ways in which they organized information, research, and globalization of honey bees.

Honey bees were taken to North America in the seventeenth century, and the next chapters study the effects of that migration on bees and women. North America defies easy organization, for its women beekeepers are some of the best documented in arguably the most complex apicultural system ever devised. The two chapters are divided based on chronology, showing in stark contrast the wealth of opportunities available for North American women beekeepers once the Langstroth movable-frame hive revolutionized the bee industry and the continent incrementally ratified women's rights to vote and own land.

The chapter on Australasia follows North American industrialization and faith-based migration patterns, although the bee industry was slow to follow other industries such as gold and cattle. Australia has only recently become a political powerhouse as varroa mites, small hive beetles, and other pathogens affect honey bees in North America. Because Australia, New Zealand, and Tasmania remained European colonies for many years, women there were shaped by European policies well into the twentieth century. Nevertheless, Australasian women beekeepers in the twentieth and twenty-first centuries have shown impressive leadership in organic honey standards and medicinal honeys.

The book concludes with the last continent to receive honey bees via human migration, South America. Since African honey bees arrived at approximately the same time as women's affirmative action policies were passed, the South American cultural and ecological landscapes remain in transition. In what was once called the "continent of tomorrow," South American women beekeepers are at the forefront of contemporary honey bee research, genetic breeding, and organic honey production.

Although some continents contain more accessible documentation than others, no continent is more important than the others in terms of its women beekeepers. In some places, women face significantly more challenges—educational, financial, political, and theological—affecting their decisions to be beekeepers. Quite often, women lead in ways outside of the hive. In those regions where women's participation in beekeeping has been veiled, ancient myths have pulsated beneath the dominant ideologies.

The rises and precipitous falls of women beekeepers make for a fascinating tale when one considers that women often have been disenfranchised of property ownership, advanced educational opportunities, and financial subsidies. I continue to hope that young women will not experience discrimination in their careers, that women will never face a dilemma between having a career and a family, and that all students will have equal access to scientific labs and well-stocked library shelves with books on such topics as physics, geology, astronomy, and poetry. These shifts will not happen tomorrow.

But while I was working in the queen production industry in Hawaii, I learned the phrase *pono pono,* literally, "right right." Loosely translated, *pono pono* means doing what it takes to resolve a situation. This book is one beekeeper's best effort to do that.

Africa
The Garden of Plenty

There are still wild places left in Africa: I can take you there.
—Liann McGregor, commercial beekeeper, 2006

A blowtorch focuses its flame on me as I walk across the windy airport tarmac: that is how Johannesburg feels. The safari leader, Keith Chisnall, asks me why I have come to South Africa. I am a beekeeper, I answer. I have an atavistic desire to be in the cradle where bees evolved.

I'm not here to see baby elephants or lions. Giraffes and zebras won't cut it. Even though I am traveling with a group of birders, I am a problem child when it comes to playing by their rules in their jeep.

The former special opportunity commander of a Rhodesian military unit surprises me. "Bees make me poop-scared!" Chisnall laughs heartily. After a very long pause, he says, "I will take you to see bees."

With Chisnall at the wheel, I slip into sleep, plain melding into plain, another measure of music unfolding one note at a time. South Africa is slow, measured, and intense. Only later do I understand that Chisnall grew up in Rhodesia, now called Zimbabwe, which is located on the equatorial line. Bees there could be called cranky. In Zimbabwe, it is not uncommon for an entire group of men to dive under jeeps when swarms zoom over mountains on their way to settle a new tree.

True to his word, the next morning Chisnall parks the jeep in front of a baobab tree with a bee nest tucked into a tree cavity. Not fifty feet away, another bee swarm has nestled in another tree stump. People are walking nonchalantly through the Kruger National Park parking lot, completely unaware that honey bees are so very close to them.

I kneel beside the bee tree, mesmerized by the bees flying into its heart. The sunlight on the scene softens the potential danger. I draw close to the bees, almost as if to fly in myself.

"Careful," Chisnall cautions, bringing me back to reality. "I cannot protect you." His camera clicks quietly. I smile, thinking that his chivalry seems quaint, but no smile returns mine.

Africa is the birthplace of bees and apiculture, inadvertently offering people ingredients—honey and beeswax—that have defined our most important rituals: birth, marriage, reciprocal payments, and gifts. For the uninitiated, there are well over twenty thousand types of bees. This book focuses on *Apis mellifera,* but even that species can overwhelm the average beekeeper. *Apis mellifera* species originated in subtropical Africa, and the continent contains at least nine subspecies: *intermissa, sahariensis, adansonii, scutellata, monticola, littorea, yeminitica (nubica), lamarckii,* and *capensis.*[1] The behavior of *Apis mellifera* species is tied to the geographic region in which it evolved.

Since Egypt has the first recorded bee histories, this chapter starts with North Africa and finishes with South Africa. Rock-art illustrations indicate that Africa's appreciation of honey bees extends much further into the past than the historic record from Egypt documents, but North Africa offers the first definitive rituals to bee goddesses, honey-hunting networks, and bee-related economies. Much later, as North Africa was colonized by European peoples, its histories included women in extension efforts with top-bar hives and modern pollination services.

Reflecting its stability and organized agriculture, Egypt was the first country to develop an organized system respecting apiculture, in which religion, daily routines, and arts were integrated by all citizens. Beekeepers used clay cylinders, which made wedding rituals easier to negotiate. "I take thee to wife," a marriage contract reads, ". . . and promise to deliver to thee yearly twelve jars of honey."[2] If only marriage rituals had remained so simple.

Egyptian wives and mothers would have appreciated honey for the same reason contemporary chefs appreciate it: honey is hygroscopic. With the use of honey, baked breads and other foodstuffs stay moist for a longer period, and in an arid Egyptian environment, this is no small feat. Equally important, honey and wax could have been used in cosmetics to make skin more comfortable and could have prevented bacteria from entering minor surface wounds.

Egyptians linked bees to the maternity goddess Nut. The sky goddess Nut was thought to give birth to Ra, the sun god, every morning. Egyptian mythology attributed the origin of the honey bee to the tears that Ra would cry. When Ra's tears fell on the ground, it was thought, honey bees emerged. Nut is the goddess responsible for music being used to encourage queen bees to emerge from their cells. Ancient Egyptian beekeepers knew that healthy hives swarmed, and they listened for the piping among queen and worker bees preceding a swarm. Because of Nut's role in the origin of bees, the ancient Egyptians made an analogy between maternal reproduction and swarming. In their word for "honey," ancient Egyptians used an image of Nut holding a reed, in effect, calling the swarm into the world. Nut was considered a nurturer, and the Egyptians knew that increasing colonies would mean eventual increases in honey yields. This goddess, and the hieroglyph for "honey," set a precedent for human societies: that if humans wanted to encourage bees, then calling or responding to the piping may assist in the swarms settling in places beekeepers preferred. As early as 404 BC, beekeepers received instructions to call queens by playing a reedlike flute. This myth must have also verified that new, younger queens could help ensure future honey production. We know now that bees do not respond to human efforts to communicate with bees by banging on pans or playing on flutes; nevertheless, the powers attributed to Nut of calling young queens or swarming bees remained a subtle but impressive belief and were adopted by other religions and cultures into the twentieth century. In fact, Eva Crane writes, "even in contemporary times, if

Piping and Swarming

In the hive, when bees prepare to swarm, they communicate in a series of sounds known as "piping." Since ancient times, beekeepers have known that queens pipe within the hive; they often connected the sounds of the queen's piping to the colony swarming. Studies of the complexities associated with queen piping began in the 1950s and revealed that queens use a variety of tones when they pipe. According to E. F. Woods (in "Queen Piping"), the purest tones emanate from virgin queens; the tones change once the queen mates and as she becomes older. They also change according to whether she is laying and the intensity of the stressors causing her to pipe. A mated queen's tone will become lower, for instance, once she ceases to lay eggs in the fall. Her tone gains in intensity when she senses the presence of virgin queens that have not emerged from their cells. The closer the virgin queen is to a mated queen, the more intense the piping will be. Sometimes, the piping is a prelude to two queens fighting for control of the hive. Sometimes, the mated queen leaves with a swarm and a virgin queen remains. On rare occasions, the mated queen and virgin queen can coexist for brief periods if they remain far enough apart.

Contemporary studies indicate that worker bees also pipe. A select group of worker bees known as scouts will pipe to alert their swarm-sisters to prepare for the flight to a new site. In order to pipe, the scout bee mounts an immobile bee; while the scout bee presses its thorax against the body of the receiver, a sound can be heard. When scout bees pipe, the swarm-sisters begin to increase the temperatures of the thorax regions in preparation for flight. The swarm-sisters are warming up their flight muscles. Worker piping and swarm warming are intricately intertwined activities. According to T. Seeley and J. Tautz (in "Worker Piping in Honey Bee Swarms and Its Role in Preparing for Liftoff"), if swarm-sisters were separated from the piping, they would not reach swarm temperatures needed for flight. Summarizing the importance of piping to the successful swarms, Seeley and Tautz emphasize that when workers pipe in preparation for swarming, their swarm-sisters respond with cooperation, trust, and timeliness.

one wants to write the word 'honey,' one draws Nut holding a reed in her hand."[3]

Nut had competition. Greek and Roman societies had their own goddesses, who in part evolved out of the Egyptian pantheon. The Christians and the Coptic Church depicted the Virgin Mary surrounded by bees, representing chastity and fertility. Nut's influence diminished considerably after the Muslim conquest of Egypt and North Africa between AD 700 and 750. Islam has provided a consistent theocratic paradigm for Egyptian women since this conquest. In the early eighth century, Islam spread across the north coastal regions of Africa, including a wide belt in the south of the Sahara, Somalia, Chad, Niger, and Nigeria and Mauritania on the west coast, and stopped in Spain. Although the Muslim religion ended in disarray after the Byzantine Empire recaptured some of its countries, Islam remains a strong mainstay in contemporary Egypt. "In 1986," writes Michael Slackman in the *New York Times,* "there was one mosque for every 6,031 Egyptians, according to government statistics. By 2005, there was one mosque for every 745 people—and the population has nearly doubled."[4]

Contemporary Egyptian marriage contracts are no longer as simple as twelve jars of honey; prospective families negotiate apartments, furniture, and steady incomes as dowries. "In Egypt and across the Middle East," writes Slackman, "many young people are being forced to put off marriage, the gateway to independence, sexual activity and societal respect."[5] Even though marriage is so expensive, "honey is still very widely used," according to British-based Bees for Development founder Nicola Bradbear. "Honey remains extremely important in weddings in many countries of Africa, and in some Muslim societies."[6]

Since Muslim religions discourage women from participating in the beeyards, few records exist of women beekeepers. Yet to the west of Egypt, Libya hosted one of the most interesting female beekeepers in all of Africa, Olive Brittan, MBE (member of the British Empire). She served as the royal beekeeper in the 1950s and lived on Mount Cyrene. Before 1952, Libyan beekeeping was primarily traditional.

The main types of honey plants were acacia, citrus, eucalyptus, and many wild plants. Because Libya was so isolated, and its bees thus had no outside contact with other bees, Brittan was convinced that she lived on the world's only "pedigreed" bee farm. She wrote in one article, "The race itself seems to be one of the purest one can find this side of the Iron Curtain. [The Libyan bee] is gentle apart from a natural intense fierceness during the great heat of the day, or when very strong through a long period of honey flows."[7]

Before Colonel Muammar el-Qaddafi took over Libya, Brittan served juniper honey to King Idris el-Senussi to soothe his throat, which was aggravated by his chain-smoking habit. She described the juniper as follows:

> When the days are warm and no north wind from the sea causes the branches of the tree to tremble, there is always a remarkable harvest to be gathered from the juniper. This honey [is] called "min" . . . and is, in fact, an exudation from the bark of the tree caused by heat. The bees cannot gather it if the wind shakes it on the ground. . . . No where on the Green Jebel are the conditions of warmth and wind so suitable as this strip of coast for so unique a source of honey—a medicament much prized by the Arabs.[8]

Other types of Libyan honey include shibrook honey from the haroob tree, min honey from the shaiee tree, and hanoon honey from the schmairy tree.

The traditional method of beekeeping meant catching swarms and placing them in long, wide, shallow boxes. These would be harvested with smoke created by lighting dung and mud, which spoiled the flavor of the honey. Furthermore, the larvae, old wax, and pollen would be mixed in with the honey. "It is not surprising," Brittan writes, "that the Royal Diwan prefers to eat the pure brilliantly-colored, delicately-flavored honey that can be extracted only from modern hives."[9]

Brittan enjoyed her time in Libya. She writes of "the old Appollonia, with its city half buried in the sea, over which one swims on calm

Empty traditional *jibahs*, Ain Babouch hives, in Tunisia. Courtesy of Ricardo Bessin.

days." Brittan also had a sense of humor in her writings. How else to interpret her rhetorical question: "Then who shall say that juniper may not yet provide a mead more potent than the all-prevailing gin, whose source is from the berry of the selfsame tree?"[10]

Sir Peter Wakefield, consul general in Libya from 1965 to 1969, explained that after Qaddafi's coup, Brittan had to leave along with the last of the British mission. But *how* Brittan left shows her British reserve. She insisted on a proper departure. Wakefield and the colonel of the mission had to "march up the hill, unfurl the Union Jack, and . . . salute. The flag was then lowered and folded, and they marched past the slumbering royal bees."[11]

A final note was written by Peter Cook, then the commanding officer of the Royal Air Force supply depot. He explained that after Qaddafi's coup, "a detachment of Royal Irish Rangers was sent on a rescue mission. They returned with a charming but reluctant and indignant lady beekeeper."[12] Brittan's papers, unfortunately, were scattered after she returned to London.

Central Africa

The central African countries such as Botswana, Zambia, and Zimbabwe offer rock-art illustrations of honey hunters that suggest women participated, even if only on the sidelines. Of the rock paintings published in *The Rock Art of Honey Hunters,* 80 percent are in Africa and 62 percent of those are in Zimbabwe.

In this region, women are portrayed as assistants because the honey bees, *Apis mellifera scutellata,* tend to be more defensive and located in tall trees or the nectar flows are inconsistent. The honey bees also have more natural predators, such as badgers, birds, and ants, and therefore react much more quickly than other types of honey bees found in temperate climates. Because of these predators, traditional bark hives are hung in tall trees. Eucalyptus trees provide a ready nectar source in the tree canopy.

Other plants in central Africa provide inconsistent honey flows. Some flower only after rain and then rapidly and unpredictably. In *Making a Beeline,* Eva Crane documented the water lilies at a mission, Maun, in the Okavango Swamp.[13] Botswana was described as "mostly bush and desert—no surface water," according to Liann McGregor. "But subterranean rivers make for unusual water features. When Angola water was redistributed to Botswana," McGregor reflected quietly, the difference was "magic, magic."[14]

These factors meant that African women did not develop a beekeeping culture as other civilizations did around hives such as skeps or log gums. Still, in the early 1950s, Zimbabwe hosted a federal agricultural program for farmers and beekeepers led by a woman named Penelope Papadopoulo. Penelope, often known as Poppy, was born in Greece and became the first female beekeeping instructor in Crete, an incredibly poor Greek island rich in nectar sources. The men initially refused to take lessons from Poppy, according to Eva Crane in *Making a Beeline.* Not to be deterred, Poppy decided to teach their wives. When their wives procured more honey from the hives than their husbands did, Poppy was finally accepted by everyone.[15]

Penelope Papadopoulo, one of the first extension specialists
in Africa. The dominant perception of African honey bees is
that they are ferocious, but reflecting the wide differences in
African bees, in this photo Papadopoulo wears only a dress and
veil. Courtesy of P. Papadopoulo.

Papadopoulo's success in Crete led to her position as senior
apiculturist at the Department of Conservation and Extension in
Zimbabwe. Being familiar with Greek movable-comb basket hives,
Papadopoulo encouraged beekeepers to use those instead of the more
expensive British and American hives, which require more wood and
maintenance. The basket hives were cheap, easily made and trans-

ported, and enabled beekeepers to manage their bees. Another nice benefit, according to Poppy's colleague R. D. Guy, was that "movable comb basket hives conserve both bees and trees."[16]

In the 1970s Papadopoulo became interested in the possibility of queen mating in enclosures, a rather sophisticated form of controlling queen breeding and genetics. Since queen bees mate on the wing, beekeepers would like the ability to limit the queens' area and ensure they only mate with certain types of drones—those from gentle colonies with good honey production, for example. Crane suggests that Papadopoulo's experiments may have been "the first to be done in the Southern Hemisphere."[17]

"Having no guidance and having read no reports by others who had tried similar experiments," Crane writes, Papadopoulo made many mistakes "through lack of experience and the results were unsatisfactory." The tower, for example, "was lined with gauze from the outside, with the result that bees became trapped between the gauze and the metal framework."[18]

Another factor was the availability of drones: "The drone colony was put into the enclosure far too soon, and queenless nuclei were left in the enclosure to build queen cells, whereas queens should have been introduced to the enclosure as young virgins. . . . Mature drones were injured when coming into contact with the gauze, and their reproductive organs everted prematurely, so they could not mate and died."[19]

Eva Crane and Poppy Papadopoulo were friends, and Crane visited Zimbabwe in 1984. In her writings, Crane shows some insights into the racial divide that continues in Zimbabwe: "I was puzzled to see very few Africans because there were eight million in the country and only half a million Europeans. But as the country we drove through became drier, I saw more and more Africans. I then realized that Europeans owned the fertile land on which profitable crops could be grown, with Africans living elsewhere."[20]

Yet, bees and honey gatherers used to do well, although women were very much marginalized. Crane writes: "Here honey harvesting

Zimbabwean woman bottling honey. Courtesy of Norman Coté on assignment for USAID.

was demonstrated—mercifully, in view of the heat, at a hive nearby; the women were astonished when I, a woman, donned a bee suit and went off into the bush with the men. I was taken to a traditional log hive supported in a low fork of a tree by the Sabi River. A smoldering piece of wood served as a smoker, and an old knife was used to cut off pieces of comb containing sealed honey, which were collected in a pail."[21]

Zimbabwe is no longer the agricultural garden it was in the 1970s. Many farmlands have been taken from white farmers and redistributed among nonfarming natives. The economy is in disarray. Inflation fluctuates almost daily. Women have fewer chances to become beekeepers, ironically, than in the 1950s.

The neighboring country Zambia encourages women beekeepers, but only gradually are the Western ideals of international aid and African realities of bee culture becoming reconciled. For years, rural development programs to Zambia have emphasized beekeeping as an economic incubator. In her exceptional book *Small-Scale Woodland-Based Enterprises with Outstanding Economic Potential*, agricultural extension agent Guni Mickels-Kokwe deals with the perception of beekeeping as a benign activity when, in fact, it may be a destructive practice. Given that Zambia has extensive forests, most honey collection—approximately 98 percent—is from bark hives. According to Mickels-Kokwe, "It has been estimated that there are some 20,000 beekeepers and 6,000 honey hunters in Zambia. At least half of the beekeepers are found in North-Western Province. Traditional bark-hive beekeeping is dominantly a male activity. Women beekeepers are few and have mostly emerged through various project interventions, e.g., to promote the use of top bar hives."[22]

Labor availability is a prohibitive factor for women beekeepers. In an interview, Mickels-Kokwe carefully explains that behind the international financial aid are assumptions that do not necessarily hold true when faced with practical beekeeping problems in Africa.

One organization, Irish Aid, for instance, provided grants for modern hives, but these grants were based on the assumption that if women are unable to go into the forest or climb trees, movable-frame box hives "fix" those challenges.[23] Because they are near the ground, however, movable-frame box hives have problems with ants. Given the swarming instinct of African honey bees, the occupancy rates in movable-frame box hives can be low. The hives have to be high to access bees. Traditional log hives, in contrast, have cylindrical shapes, and since they are smaller and lighter, they can be hung or placed on branches in a tree. Beekeepers want to place hives higher to make it easier to attract bees at their usual flying height and to protect hives against fire, honey badgers, red ants, and other pests.

Another development program, International Fund for Agricultural Development from Rome, Italy, focused on a forest resource management project completed in June 2007. It emphasized "non-wood products" in an effort to stop deforestation. "Beekeeping was the most successful project of all," summarizes Mickels-Kokwe.[24] The consequences of deforestation on beekeeping potential in Zambia are far more serious than those of bark-hive harvesting.

So, in Mickels-Kokwe's opinion, African women can be successful at beekeeping as long as they are permitted to hire out some of the labor to place and collect honey from bark hives, but this one factor stymies many aid agencies that want women themselves to become beekeepers. Mickels-Kokwe's point is important: women do not have to be beekeepers to profit from bees. They can barter with honey beer. Honey brewery is a full-time occupation, and it is a way of raising labor. Food and beer can be used to pay men while providing social entertainment and cohesion. Honey beer is a valuable commodity for trade and is more suitable for the less urban settings. Table honey, in contrast, is a more refined product and has mostly urban, middle-class demand. A relatively recent demand has come from people with HIV, who eat honey for its perceived healing properties.[25]

In Northern Zambia, assumptions are not very fixed. "When confronted with proof, ideas can change, the people can be flexible," Mickels-Kokwe explains.[26] She offers some perceptive insights into the gender norms in Zambia:

> There is no restriction on female beekeeping, *per se*. Men perceive women as being constrained by the fact that hives need to be hung in trees in remote places in the forest. Women in general are not comfortable about climbing trees. It was also considered impossible for them to leave the homestead chores and go and camp in the forest. . . . Women's participation is further dependent on the husband's permission. The role of the husband in decision-making is quite pronounced in rural households in North-Western Province. It will therefore be very difficult for a woman to decide to take up beekeeping and start employing labourers to tend to hives without the active support from the husband.[27]

Mickels-Kokwe reminds her readers of some capitalism basics at work in Zambia: "It is not uncommon to find that households work together as a production unit, drawing upon the labor of many members of the household." However, she explains, "when it comes to the marketing sales these are the individualistic, with the husband and wife managing the sales from their individual fields. A beekeeper who cannot consider sharing financial responsibility with his spouse is not likely to do so with any other person."[28]

In short, Mickels-Kokwe concludes that Zambian women are flexible to new ideas. But when international aid societies impose inflexible standards, such as requiring women to use the Langstroth movable-frame hives or requiring women to be beekeepers, African women do not prosper because those standards are based on assumptions that often do not match the land-based conditions such as high trees, predators, and climate.

Similar to Zambia, Kenya is an African jewel. It differs from other African countries because of its peacetime economy based on coffee, flowers, and tourists. Even in the prehistoric past, the landscape remained comparatively stable, with tropical forests being relatively constant features as opposed to the shrinking savannahs in other parts of Africa. These immense forests once again mean that native women are sideline participants in honey gathering. Authors David Brokensha, H. S. K. Mwaniki, and Bernard Riley, writing in the 1970s, commented about the gender restrictions: "Women are alleged to be unable to endure the hazardous journeys and tree-climbing."[29]

Brokensha, Mwaniki, and Riley also provided a picture of Kenyan resistance to government reservations, specifically because it would have meant a move away from bee forage. The people refused to move: "The Mbere division is a large, nearly empty *Acacia-Commiphora* dry savanna near the Tana River and is suitable for beekeeping. When the government tried to move them in 1915, the Mbeere refused, in part, because they didn't want to leave beekeeping."[30]

Kenya became independent in 1963 but lagged behind in apiculture. Writing in 1984, K. I. Kigatiira described traditional beekeeping. At that time, it was "restricted to individual males who inherit techniques from their fathers. They keep a number of horizontal cylindrical hives made from tree bark or hollowed-out logs, and these are cheaper than any improved hive."[31] But the Kenyan honey collectors used artifacts that suggest feminine artwork such as goatskin bags used to gather and carry honey and wax.

Documenting how quickly gender divisions can change, Margaret Adey in 1985 wrote about the Kibwezi Women's Pride, a beekeeping women's cooperative with a membership of two thousand and a regular honey trade network established in 1981. As in other African countries, the cooperative used Kenyan top-bar hives instead of movable-frame hives. Top-bar hives offer the advantages of sustainability, low cost, and considerable variation. Furthermore, in a top-bar hive, beekeepers use division boards, which allow more flexibility as bees increase. The Kibwezi Women's Pride used Canadian assistance to get started, but "members have built their own refinery with homemade bricks and each month they process 600–1000kg of honey."[32]

Adey does not say a word about honey beer, but other visitors to Kenya remark at great length about the honey brewery industry. Brokensha and his colleagues mention "one bee-valley [that] is famous for an area in it known as Kauriro, where the bees and honey are so abundant that some collectors are said to 'got lost' from drinking too much honey beer."[33]

These authors also provide a few examples of honey songs, which number about fifty; the songs are "musically interesting and sociologically illuminating, but were often condemned by church groups because of their ribald humor and their connection with honey beer."[34] Brokensha and his friends did not provide specific examples of ribald humor.

Even Eva Crane, the paradigm of British propriety, went to a beer hall. Not just one but three. When writing about her adventures in *Making a Beeline* much later, she noted that in 1967, "this provided for great excitement for the local people: a white woman rarely went to Marimanti, let alone into a beer hall."[35]

Crane provided the following details just so we can visualize such a place:

> Each hall contained one or two 180-litre casks holding 90 litres. Everything collected from the hive, combs, pollen, honey, and any adhering bees, was put into each vessel with water. A piece of pith from the gourd *Kigelia Africana,* which carried the necessary yeast, was floated on top of the liquid and moved from one cask to the next. When all the

African Honey Beer

Should readers feel inclined to try their own version, Brokensha, Mwaniki, and Riley offer a recipe for honey beer: "[Beer is made] by placing comb in water and rubbing it by hand. After rubbing, the comb is set aside to be used in perfuming a hive. The solution is placed in a gourd, one with a wide mouth and short neck to which fruit is added from the sausage tree muratina *kigelia aethiopium* as a fermenting agent. Gourds are allowed to stand in the sun, or near a fire, for twelve hours, and the liquid is ready to drink" ("Beekeeping in the Embu District, Kenya," 121).

honey was used up, the pith was dried on a hut roof and kept until the next honey harvest. Liquid in the vessels was bubbling away, and customers were served by dipping a *bakuli* into the brew; a *bakuli* was a half-gourd or an aluminum basin. The beer was expensive: a *bakuli* of it cost a quarter of a farm labourer's daily wage.[36]

In traditional society, only old women and elders were allowed to drink beer, but Crane acknowledged that many younger people drink in contemporary societies.

Honey beer is one of the reasons why men buy honey in Christina Kessler's sublime retelling of an old Ethiopian tale, *The Best Beekeeper in Lalibela*. Although the backdrop to this story is King Lalibela's ability to attract swarms as an infant, he is not the protagonist. A young girl, Almaz, decides that she wants to be a beekeeper. Almaz is determined to outdo the honey sellers in the market, who are all men. She intends to have her own hives.

All of the men laugh at her, except for a young priest, Father Haile Kirros, who encourages Almaz to follow her heart. Almaz makes a hive and provides gorgeous honey at the market—until one day the ants have eaten the honey. Once again, Almaz must figure out how to be a beekeeper using nontraditional methods. Father Haile Kirros again coaches her: "It [beekeeping] is great, and all great endeavors require great effort. Who knows, the answer could be something very simple."[37]

Almaz places bowls of water under all four legs of the hive stand. Then the ants cannot climb into the beehives. Almaz even conquers her fear of getting stung, working the bees "with a steady hand and a clear heart."[38] The lessons of this book are very clear to all young people: great accomplishments require great effort and simple solutions.

South Africa

Africa has been a honey-hunting continent for centuries, but nineteenth- and twentieth-century colonization by the Belgian, English, French, and German governments meant that more white women became involved with beekeeping. These women often had to go to Europe for training and adapt that training to their own resources, but they were as successful as their male counterparts in spite of the difficulties with African bee management.

The first records of European women beekeepers in South Africa occur as early as 1875. On November 3, 1875, a person named Mrs. Mullens ordered a stock of Ligurian (*Apis mellifera ligustica*— Italian) bees from Neighbor and Sons to be shipped to her address in Grahamstown. Not only did the bees have to survive a sea journey but they had another two days' journey by wagon once they arrived. According to Peter Barrett, "They were released from the hive on September 3 and appeared weak at first but began to work in less than an hour. A large number of dead bees were found at the bottom of the hive on opening—most likely caused by the boat in which the bees were, having water in it."[39] Nothing more is known of Mrs. Mullens, why she was in South Africa, or how the bees fared.

"FIENDS INCARNATE"

In a 1917 publication titled *Beekeeping in South Africa: A Book for Beginners,* author Alfred Attridge encouraged women to keep bees. However, his remarks seem to apply to European beekeeping, not necessarily beekeeping in South Africa. Nor does his advice acknowledge the differences in African honey bees or urge caution in terms of dress and hive movement. For instance, he is cognizant that there is a lot of lifting and carrying of heavy hives. His solution—"one method is to get a man to do the lifting and carrying"—was not always practical or feasible.[40]

Europeans brought their cultural norms with them to Africa as well as their European bees. An article from the 1930s focused on a Miss A. Clayton who kept bees in the Eastern Transvaal. Miss Clayton complained not of the bees but of thieves: "Recently thieves, Natives, not only stole honey but did a great deal of damage to the colonies by cutting out combs from the frames and completely spoiling what they did not carry away. One thief was caught in the act but got off with the comparatively light sentence of a month's imprisonment."[41]

The author of the article offered his full support, noting that when thieves damage hives, they often damage the queen and brood in the process of taking honey. Another author, F. H. Cooper, advised beekeepers that a "readiness to sell dark honey cheaply to the laborers on farms has probably done as much as anything to check thieving."[42] Even today, these same problems continue. Contemporary South African beekeeper Liann McGregor complains of stealing but notes that the thief "tends to be familiar every year."[43]

Miss Clayton learned beekeeping from a Miss Pullinger, who was from Elgin, Cape Province, and had studied queen production in England with Mr. Herrod-Hempsall. Clayton bought Miss Pullinger's equipment and four years later had between 125 and 175 hives to pollinate citrus.

When asked about the temperament of her hives, Clayton remarked that the bees were "fiends incarnate" and "will sting through anything" at the end of a honey season.[44] Clayton started taking honey off during the night after the staff tried wearing two pairs of riding breeches, a shirt and a long coat, gloves, leggings, boots, and wire veils. But she was careful to specify that the bees were cranky only during October. The article ends with Clayton's failures to interbreed an Italian queen with African drones to make the colonies gentle.

"TEMPERAMENTAL TOURISTS"

In the 1950s, *Farmer's Weekly* published an article about a couple who started a pollination business during the Depression in the 1930s. The Gouses took bees from the Southern Transvaal to the blue gums in the Rand and finished among the orange and mango orchards in Tzaneen and Duilvelskloof.

Mrs. Gous and her husband were blessed with a nine-month honey flow, which enabled them to sell clover and orange honey to customers throughout the Depression. Nevertheless, they suffered the dangers associated with migratory beekeeping: "Once the Gous couple was taking a load of hives to new nectar fields when the brakes of the truck failed on a hill. They and a couple of hundred thousand infuriated bees piled into a donga [a ditch in the road] with the wreckage of the truck on top of them. But man and wife, instinctively alert to any emergency, got out in double quick time to run for their lives."[45]

Contemporary beekeeper Liann McGregor explained how she managed to maintain a commercial business in South Africa for forty years. She and her husband, Lloyd, wanted their children to grow up in a rural environment since she herself had grown up in the African bush. So she started keeping bees, which facilitated the rearing of their children. They had 30 hives at first and increased to 450. Now they have about 250 hives.

South Africa is the Garden of Plenty, but the areas vary widely and have differing types of rainfall. In 1965 the McGregors started offering pollination services to the orchard growers in the Free State, pollinating cherries, apples, pears, and plums. In the southwest, near Cedelia Hills, the McGregors gleaned prolific honey harvests in the 1970s. Natal receives forty inches of relief rain and ocean winds on coastal plains, an effect Liann McGregor called, poetically, "shedding the rainfall." The rain is steady and continuous. But Drakensburg, where many prehistoric rock-art illustrations are located, receives convectional rain. Evaporation and formation of clouds mean more erratic rains and storm conditions.

The different conditions in the various South African landscapes require different beekeeping techniques. When the temperature drops to five degrees centigrade, there is often only one honey flow from January to May—that of eucalyptus. Asked whether her children were encouraged to go into beekeeping, McGregor answered in the negative: "There is only income five months instead of twelve months. Seven to eight months of no income is very difficult."[46]

Furthermore, the Cape bee *Apis mellifera capensis,* a race that developed in South Africa, is a real problem. In some circumstances, workers as well as the queen can lay eggs that develop into full females. The Cape bees take over honey bee colonies by "posing" as honey bees, eating the honey, and leaving the hives empty. The Cape bees are not honey bees and do no gathering of nectar and pollen, thus colonies collapse in weeks.

In talking about the proper maintenance needed for her beeyards, McGregor offered abundant information. Never go back to a hive that has already been worked, for starters. The pheromones have already been activated, and by going back, a beekeeper just causes more agitation.

McGregor learned a lot from the cattle industry when she first started since there wasn't a bee extension service and a serious "good ol' boys network" was not always helpful. By going to cattle confer-

A Commercial Beekeeper's Field Tips

Liann McGregor had these points to offer in terms of working with *Apis mellifera scutellata:*

- *A. m. scutellata* bees are not as vicious as they seem.
- They have more similarities than differences with other bees.
- Proper management of *A. m. scutellata* "boils down to knowing how to handle them" and "not taking them for granted."
- Smoke the bees for a longer time before entering an *A. m. scutellata* hive.
- Do not return to an *A. m. scutellata* hive after it has been worked.
- "Taking a whiz while working African honey bees is difficult."

ences, she learned to transfer good management practices from the cattle industry to her own. If one has cattle, for instance, one has to dip them. Ditto the bees. If one has bees, one needs to maintain them.

In time, the government would implement a bee extension service, and integrated pest management would become the norm. But for the longest time, McGregor had to train herself to realize that the most important aspect of maintenance was "not what the eye sees, but what it doesn't see."

Because McGregor was a beekeeper throughout apartheid, she offers a perspective on the rural beekeeping development programs that few others have. One government indemnity program ended because beekeepers abused the benefits that came with burning hives that had foulbrood. Once apartheid was over, the extension programs at Pretoria fell away with the change in government. So there are no inspectors. Combine that lack of oversight with the lack of an extension service, and a beekeeper can still feel hindered by a "good ol' boys network" system. And perhaps another cultural condition is a factor too: "Beekeeping in South Africa is still considered a *bayvoeur* or 'peasant's job,'" Liann explains. "If nothing else, the thinking goes, you can always be a beekeeper."

One government program, Black Empowerment Employment (BEE), is the object of satirical cartoons. The idea behind BEE was that beekeepers who could not make it commercially could be taught how to be sideliners. Black women were more amenable to beekeeping than black men, but cultural considerations and political challenges make beekeeping more difficult than many people realize.

As a child growing up in the South African bush, McGregor learned mechanics when she was fourteen. She had a father who was an accountant. "You have a brain," he told Liann. "Use it." Liann builds her own hives, which saves money. Her husband, Lloyd, who taught for twenty years, is also progressive in his role at home. He runs the honey extraction part of their business. "I never get sticky," Liann smiles.

McGregor is skeptical that a woman can be as successful a beekeeper as she has been simply because the technology is so advanced.

The differences between success and failure are in the details. The best way to describe beekeeping economics in South Africa is "linear." "Look around you," she says, "and see what else is produced already."

McGregor's long-time male employee is of Zulu heritage. There are two cultural considerations that have to be factored into their discussions: courtesy and time. The first consideration can lead to all kinds of misinterpretations, because answers to questions are misleading in an effort to be respectful. If there is a problem with the hives, it may go unreported out of respect for Liann.

The second consideration means that her employee often does not have a contingency approach to a situation. In terms of maintenance for the vehicles, for example, it is a high priority to change oil and do other basic preventive care. But those are tasks that Liann must insist on, not expect her employee to remember to do. Because of her cultural upbringing, McGregor has learned a valuable commandment: "Discipline sets you free; it is intrinsically related to time." But as far as her employee is concerned, "There is no tomorrow." In the Zulu world, there is a saying: "Where will the sun be when I arrive, not how far is it."

Not only are vehicles today more sophisticated for the beginning beekeeper to learn to maintain, but environmental programs such as Working for Water advocate that all alien vegetation should be cleared. The problem is that "alien vegetation" can sometimes mean the deciduous fruit industry, which depends on pollination. McGregor, of course, depends on those orchards' pollination fees. "When a species is declared 'alien,'" she explains, "a honey producer can go out of business because the crop can be dependent upon that flower."

In summary, McGregor offers the following thoughts about her life as a beekeeper. For the first half of her career, it was commercially viable. But lately, theft and other problems have eroded the profit margins. Challenges such as diseases, predators, and thefts are surmountable, but her profit depends on keeping a skeleton crew, making equipment, and servicing her own vehicles. She works fourteen-hour days and fortunately has a supportive home environment. "After forty years," she says finally with a smile, "I am still not bored."

In rereading Eva Crane's experiences in South Africa in *Making a Beeline,* I feel fortunate to have visited South Africa without apartheid. South Africa at times felt more like Alabama than a foreign country, but it is a country with many freedoms.

By contrast, Crane noticed in 1973 that "what was known as 'petty apartheid' caused continual resentment: the use of separate footpaths across railway bridges, separate doors to post offices, separate park seats and separate beaches."[47] When Crane visited an Afrikaner beekeeper, the only women involved with the business were girls sticking labels on jars. On the weekends, "these laborers were bussed to the outlying township where their families lived." Her efforts to try to establish a friendship "seemed to cause embarrassment. We drove back to Pretoria in the rush hour, among shabby 'non-white' buses and nicer 'white' ones, 'black' queues, post offices with 'white entrances' and 'black' entrances and so on. I felt a long way from Ethiopia." Still, Crane subtly commented, "there was only one sea for them to bathe in."[48]

While in Africa, I learned one word that was more important than all the others: *fungwe.* The closest equivalent in English is "desire," but in African languages, *fungwe* is a deeper thirst and instinct to culminate an attraction for something or someone; a desire that, if fulfilled, will shape one's entire destiny.[49]

Fungwe, for me, is defined by honey bees. Honey bees bring me closer to the pulsating life force driving our evolutionary fits and starts than I have ever been. To go to Africa is to know much more than the differences in hives (from clay cylinders, to log hives, to top bar and movable baskets) or even the differences in subspecies (from *A. m. adansonii* to *A. m. scutellata* to *A. m. lamarckii*). In following honey bees and those who care for them, I am led to a deeper awareness of those who live in modest circumstances and are marginalized by social conditions. I am struck by the loss of opportunity but humbled by the hospitality shown to me and the kind invitation to return from Liann McGregor: "There are still wild places left in Africa: I can take you there."

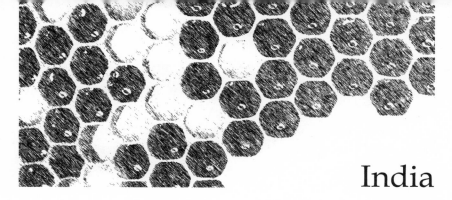

India

The Heart of the World

Behold! The kwehhrshy flowers have lit up the thickets. Look!
The queen bees are sucking their nectar.
—Toda chant

India's plethora of honey bees, its honey-hunting history, and its diverse faith-based religions have imprinted women primarily through artistic, literary, and femininity rituals such as bridal and maternal rites of passage. There are four overlapping areas in which women have been shaped by or are in the process of shaping India's contemporary bee culture: the honey-hunting cultures; the Hindu and Buddhist religious rituals; the British practices and assumptions under colonialism; and emerging opportunities for India's women beekeepers in a new global economy.

Adivasi Customs

A number of adivasi native populations, such as the Badaga, Kurumba, Kota, and Toda peoples, still practice traditional honey-gathering techniques. Their literature documents how women celebrate femininity using honey. The Toda women are known for their embroidery. In an excellent book by Kunal Sharma and Snehlata Nath, *Honey Trails in the Blue Mountains,* the authors report that "one of the oldest motifs is the *Kwudrkorr pukhoor* or the honey comb pattern. This depicts the inside of the honeycomb and is sometimes even embroidered with the brood shown within! Its importance lies in the fact that this was done traditionally only on the cloak of the departed so as to ensure a safe passage to the after world."[1]

Unlike other honey-hunting cultures in Nepal, the Wynaad indigenous communities of India collect honey as a group. According to Sharma and Nath, "During the honey season all people of the village go for honey collection—men, women, and the young." It seems the Wynaad children are like mischievous children everywhere: Sharma and Nath write, almost as an aside, "Children are strictly monitored as they may eat lots of honey leading to low amounts for sale."[2]

Even though most activities of traditional honey collection are male oriented, it would be a mistake to see all the native peoples as paternalistic. The Kurichiars people "are widely considered as one of the earliest inhabitants of the Wynaad hills. They have a vibrant history and had revolted against the British with the forces of the Pazzhasi Raja." They are exceptional honey hunters. More to the point, Sharma and Nath explain, "they are the only matrilinear society in South India. The women participate in agricultural operations, fishing, animal husbandry, fuel collection, etc. Most land is owned by lineages, whereas there are few individual owners in present times."[3]

The Badaga people also have deep roots in the Nilgiri region, their language being the lingua franca there before the arrival of the British. "They were supposed to have come to the Nilgiris after the break up of the Vijaynagar Empire in 1565," according to Sharma and Nath.[4] When the empire was taken over by Muslim forces, the Badagas settled in the area for nearly two hundred years, becoming agriculturists. When the British arrived in the Nilgiri territory in the early 1800s, a medical doctor and botanist, Francis Buchanan, meticulously recorded daily practices and habits in his journal. He had been assigned by Marquess Wellesley, the governor general, to investigate the state of agriculture, commerce, and development in the dominions of the rajah of Mysore acquired by the East India Company. In Buchanan's journal, which began in 1800, he referred to the Badaga people as "honey and wax gatherers." Buchanan did not seem to realize that for a long time before the British arrival, the Badagas had been called the *madhura* (honey) or *maanthakula* people.[5]

The Badagas had a honey goddess called Kunnimaara. "She was initially a Kurumba goddess," Sharma told me in correspondence, "but was later initiated into the Badaga pantheon."[6]

The Kurumba people also attach maternal privilege to the rocks that provide shelter for *Apis dorsata*. Sharma and Nath explain, "They believe that the rock is their 'mother,' the rope is their 'father' and the tree on top of the cliff is their anna, or elder brother. After honey is harvested, they keep some on the rock, and under the tree first, before eating it themselves."[7]

Hinduism and Buddhism

The more conventional cultures in India—Hinduism and Buddhism—have clearly defined maternal and bridal rituals using honey. Even the Hindu god Kama, the Indian god of love, flies with a bow whose string is made of bees. Hinduism began around 3000 BC with the collection of ancient religious stories called the Veda, gathered by the Aryan people in the Caucasus region. There are four texts in the Veda: the Rigveda, the Samaveda, the Yajurveda, and the Atharvaveda. The primary text for Hindu people is the Rigveda, which first codified courtship practices, marriage rituals, and birth ceremonies. "The *Rig-Veda* was the first of their sacred books," explains Eva Crane in *The World History of Beekeeping and Honey Hunting*. It was "accumulated over centuries—probably both before and after the migration. Around 1000 B.C. these were collected together."[8]

In the Atharvaveda, a charm in which licorice is used to secure a woman's love reads as follows: "This plant is born of honey, with honey do we dig for thee. Of honey art thou begotten, do thou make us full of honey. At the top of my finger may I have honey, my tongue's

Honey Ceremonies

Scholars tend to think the Rigveda was in compilation before Aryans migrated to the Caucasus region because the word for honey, *madhu*, is etymologically identical with the Greek *methu*. In *Honeybee: Lessons from an Accidental Beekeeper*, Marina Marchese notes that "the first syllables of *madhuparka* mean honey; the Sanskrit word *madhu* and the Chinese word *myit* are related to the words of *mit* of the Indo-Europeans, *medhu* of the Slavs and *mead* of the English" (99). *Madhuparka* is a sweet liquor made from sugar, ghee, curds, herbs, and honey. It is given to guests, to a suitor proposing marriage, and to pregnant women in their advanced trimesters. Newborn sons are also allowed to taste it in India, although this is against the law in the United States because of the risk of infant botulism.

Another ritual prefiguring the birth of a child is when the native tribe, the Toda, incorporates the bow and arrow in its paternity rites. A man ceremonially gives a bow and arrow to his wife during her seventh month of pregnancy, symbolizing his commitment to protect the child to be born. At the end of this ceremony, the woman is expected to place some foodstuff at the base of the *kaihhrsh Eugenia arnottiana* tree, including a piece of honeycomb. This practice is now subject to availability (Sharma and Nath, *Honey Trails in the Blue Mountains*, 258).

root sweetness of honey. Sweet as honey is my entrance, sweet as honey my departure. With my voice do I speak sweet as honey, may I become like honey!" The marriage ceremony itself involves a ritual in which a husband ties the medhuka flower to his bride's body, singing, "Full of honey the herbs."[9]

The highlight of the Hindu marriage ceremony is *Saptapadi*, the Seven Steps. With the bride's sari wrapped around the groom's *kurti* (tunic), the groom leads his bride seven steps around the fire. The altar moment transforms the bride and groom into god and goddess in human form. In many parts of India, the bride is considered Lakshmi, the goddess of fortune, and the groom is her consort Vishnu, the Great Preserver.

In North Indian weddings, after completing the Seven Steps the bride and groom say the following:

We have taken the Seven Steps. You have become mine forever. Yes, we have become partners. I have become yours. Hereafter, I cannot live without you. Do not live without me. Let us share the joys. We are word and meaning, united. You are thought and I am sound. May the night be honey-sweet for us. May the morning be honey-sweet for us. May the earth be honey-sweet for us. May the heavens be honey-sweet for us. May the cows yield us honey-sweet milk. As the heavens are stable, as the earth is stable, as the mountains are stable, as the whole universe is stable, so may our unions be permanently settled.[10]

With each step, the bride and groom throw small bits of puffed rice into the fire, representing prosperity in their new life together. This act seals the commitment between the two.

Another marriage ritual involves the honeymoon night, in which a bride's vagina is smeared with honey. During the process, an erotic prayer is chanted: "Damsel! I anoint this thy generative organ with honey, because it is the second mouth of the creator: by that thou subduest all males, though unsubdued; by that thou art lively, and dost hold dominion."[11]

Before the twenty-first century, mothers used to mix their milk with honey to give their newborn babies. Among the Hindu in Punjab, honey and sweetmeats are often passed over children's heads to drive evil spirits away. Honey is also used to moisten the lips of the newly born first son. Other social practices include giving honey to women who are five months pregnant and giving honey to suitors about to ask for the hand of a young woman.

So, in the course of a Hindu woman's life, her diet would depend on five sacred nectars: milk, honey, water, curd, and butter. Not only providing nutritious content, these items would nurture her through major life-changing stages—marriage and pregnancy, being a wife and community participant—with honey bringing all the transitions together.

Whereas the Hindu religion is so ancient that no individual prophet is defined, the Buddhist religion began with one man, Siddhartha Gautama, a Sakyan prince. His quest to find happiness through spiritual guidance rather than material gains has provided simpler ideologies for women and bees. The Buddhist dharma is a code by which adherents seek to leave worldly responsibilities behind in an effort to perform good deeds. The rationale is that the person will be rewarded in the next life for years of service in his or her current life span.

Honey is an important part of almsgiving throughout the year because Buddhist monks and nuns have very little to eat. During the honey festival, held during a time of honey scarcity, the monks and nuns set out bowls for honey collection.[12]

Although honey is an important symbol and dietary staple, Buddhist and Hindu women traditionally have not served as beekeepers. Beekeeping was considered a "low-caste" occupation. In fact, in the caste-centric Hindu religion, Brahmans, who were in the highest level, could own bees but not actually work them.

In *Asian Honey Bees,* Benjamin Oldroyd and Siriwat Wongsiri note: "The widespread influence of Buddhism and Hinduism meant that technical advances in agriculture, including bee hunting and beekeeping, tended to spread more quickly within Asia than they did in other parts of the world such as Africa, Australia, and the Americas, where human populations were more dispersed and less communicative." They speculate that "this may account for the uniformity of honey-hunting techniques and traditions across much of Asia."[13]

British Colonialism

Feminine honey rituals in India stayed intact for centuries until the British arrived at the turn of the nineteenth century and immediately began to affect the ancient, forest-based Indian bee cultures by deforesting huge tracts of land and introducing new ideas about beekeeping and honey bees. "Green gold" is the phrase that the British used to describe the tea, spice, and trees when they arrived in 1799. British colonialism altered honey-hunting patterns and native tribal roles because of the massive deforestation that happened in three major forests, the Assam, the Nilgiri, and the Wynaard. According to forest historian Michael Williams, "These areas owed their existence to rapid accessibility and cheap and efficient transportation links with the wider world: Assam for tea, the Nilgiri Hills and Wynaard Plateau of Kerala for coffee and subsequently for tea, and Sri Lanka, formerly Ceylon, for coffee and tea, the stimulants that were exported to satisfy Europe's continually expanding 'soft drug' culture."[14]

As with the British colonists who came to North America, Williams reports, "there was an animosity toward uncultivated land, which was regarded as 'waste.'" This attitude toward wilderness was found throughout Europe during this period. "Forests were perceived as the abode of the unruly and disorderly, from the murderous thuggees down to the run-of-the-mill thieves," explains Williams, "so that the elimination of the forests would mean an end to lawlessness, as well as unproductive land and population."[15]

For almost eighty years, very little appreciation or acknowledgement was given to the forest-based honey-hunting tribes in India. The first glimmer of British interest in India's apiculture occurred when the colonists transferred Italian honey bees to India in 1882. Nineteenth-century government beekeeping manuals written in 1884 derided indigenous systems of beekeeping as "barbarous to the bees, less productive, precarious, and not admitting to expansion beyond the limits of an industry to be pursued by the peasantry on a very small scale," to quote British bee inspector J. C. Douglas from the *Handbook of Beekeeping for India*. Douglas optimistically declares, "Beekeeping in India has before it a great future; as in the other economic arts it is for the European to lead. When the benefits to be derived have been demonstrated, no doubt the natives of India will follow."[16]

Douglas was sincerely concerned about poor women and saw beekeeping as a way to improve the quality of life in rural India. Using his observations of European women beekeepers, Douglas explains: "It is a pursuit in which women readily find pleasure and profitable occupation, many having been very successful in England and America."[17] Addressing the fear of stings, he assures women that

A Nepali woman named Januka, opening a hive for the first time, in a village named Syanje at the base of the Annapurna range. Courtesy of Ed Levi on assignment for USAID.

with time, their awkwardness will subside and there will be fewer stings.

Since India did not have a bee industry or commercial bee equipment readily available, Douglas advised his readers to adapt the European beekeeping patterns to local Indian customs, "labor being cheaper than machinery." Douglas even advises his readers to appreciate the native bees, acknowledging an English lady who had decided to write about them but not mentioning her name.[18]

J. C. Douglas's book was followed by another written during the twentieth century by imperial entomologist T. Bainbridge Fletcher, who was less optimistic about beekeeping in India. He does not include women in his analysis published in 1915, although he stays gender neutral. He states simply: "Beekeeping cannot be looked on as a source of income in India except perhaps in a few exceptional localities and even then only in cases where the beekeeper is a practical expert willing and able to take the necessary trouble to attain results."[19]

His advice regarding those pursuing a lucrative career in honey is still true: "Beekeeping is not a short cut to fortune anywhere, probably less so in India than in most other countries." His reasons: the difficulty of getting honey to market, the lack of technology, the difficult terrain, the shortage of labor. All of these circumstances were enough for him to discourage any dreams of making the hobby profitable.

Emerging Opportunities for Indian Women

If women in India kept bees in movable-frame hives during the early twentieth century, the records are sparse. "Efforts to keep bees in wooden hives probably started at the time of the Gandhian self-reliant movement in the 1940s," according to Md. Nurul Islam in "Beekeeping in Bangladesh." "Prior to this people kept bees in logs, clay pots, or similar methods."[20]

Movable-frame hive beekeeping was introduced to Karnataka during the latter part of the twentieth century, according to Bees for

A married woman stands beside her kitchen hive. The
bees are *apis florae,* the tiniest of Asian honey bees.
Courtesy of Ed Levi on assignment for USAID.

Development founder Nicola Bradbear. "Ninety colonies of *A.mellifera*
were imported in March 1996. Some of these died, and those surviv-
ing were in a weak state. None has so far generated honey surpluses
greater than can be obtained from *A.cerana.*"[21] After so many years of
British colonial rule, Nicola Bradbear and M. S. Reddy suggest, there
is still much to be learned in international beekeeping programs. In
their article "Sustainable Beekeeping Development in Karnataka,"
Bradbear and Reddy hint at the complexities regarding apiculture
among ancient honey-hunting cultures: "Honey hunters, who collect
from the wild, are barely regarded by extension staff as part of the
'apicultural community.' In interviews, honey hunters said they received no assistance from the extension service."[22]

In the 1990s, according to Bradbear and Reddy, there was "no
strong request arising from beekeepers to introduce *Apis mellifera mellifera* [German or, more generally, a European honey bee]. . . . Some
beekeepers have observed that it does not thrive. They have noticed
that during the rainy period it ceases foraging while *A.cerana* continues, and also that it is more susceptible to predators."[23]

As India emerges as a global powerhouse in the twenty-first

century, many international groups have made concentrated efforts to include Indian women in beekeeping workshops. Arkansas apiary specialist Ed Levi travels to India and Nepal with philanthropic organizations such as Winrock to teach basic beekeeping to women, using movable-frame hives. However, Levi prefers that the group use *Apis cerana* bees if possible, especially at higher altitudes. He has found women and children very receptive to beekeeping. More than a few women maintain hives in their kitchens, although few make this type of beekeeping a commercial industry because bees do not make much honey. The extraction process is also primitive. The people who sell their honey at the market command a good price because buyers believe in the strong medicinal value of honey.[24]

Levi has also taught beekeeping to the Boté people, nomadic folk who live along rivers and have been disenfranchised from the property-owning society around them. Given their nomadic lifestyle, the Boté people were taught to catch swarms. But the Boté women can also adapt to capitalistic ways. In one lesson well learned, a Boté woman caught a swarm and promptly sold it back to Levi.

In addition to the workshops provided by Winrock, the Darwin Initiative began in the Nilgiri Biosphere Reserve in 2006. This three-year project focused on four main components: research about *Apis dorsata* and local tribes; infrastructure among field centers, trained staff, and forestry officials; dissemination; and advocacy. In all four components, women defined new roles. The Darwin Initiative came at a time when women were increasingly having more say in land-ownership. Even though women have the right to vote when they turn eighteen, they have not necessarily enjoyed the right to own land. "Traditionally, women did not have the rights to land, especially because the sons used to be handed over the deed to their ancestral property," explains Sharma. "Now, however, things are opening up and positive legislation is bringing some semblance of rights for women, though the first right for land is still with the men."[25]

Agricultural programs at universities are establishing programs for women. Tamil Nadu Agricultural University inaugurated a new

This Boté woman wears a veil. The Boté people are nomadic "river people." Beekeeping is part of an economic
development plan to teach agrarian skills. Courtesy of
Ed Levi on assignment for USAID.

beekeeping program for women in 2007. Dean of agriculture K.
Ramamoorthy explains: "The mandate of Krishi Vigyan Kendra is
to see that farmers learn about allied activities of agriculture. Since
the demand for honey is good and expanding, this scheme was
launched."[26] The training sessions last for three days, and each work-
shop is financed by the Department of Biotechnology in New Delhi.
"We chose this trade because requirements for pure honey are increasing in drug/pharmaceutical companies and retail stores," says
Dean Ramamoorthy.

A woman checking her hives in a village named Syanje at the base of the
Annapurna range. Courtesy of Ed Levi on assignment for USAID.

Sharma also explains: "Women beekeepers are traditionally known to be amongst the hill communities of the Western Himalayas, especially as the men in Uttaranchal move down to the plains. During that phase, it is the women who manage apiaries. So, there is no definitive date for women to have entered into beekeeping, but they are known to be present in at least small numbers throughout history."[27]

Since India has always placed a high value on medicinal treatments with honey and bee stings, North American beekeeper and apitherapist Reyah Carlson traveled to India to offer bee stings as relief for medical conditions. Her trip to Vadodara Baroda in the state of Gujarat was arranged to assist a woman suffering from multiple sclerosis. Carlson's travelogue of her trip is more than mere diary. It reflects strong familial ties, faith, and hospitality, with the honey bee as an international vehicle for optimism, therapy, and ambassadorship. Carlson describes her initial impression of India as one of exhaustion immediately replaced with "pure adrenaline." The rest of her journey reflects an immediacy to India and the environment; her blog details some of the rewards and challenges she had in offering apitherapy.[28]

Carlson's experience as an apitherapist blends medical philosophies between East and West. The familial networks that were strengthened by her work suggest previous ceremonies from centuries past. Perhaps because of the small numbers of the native adivasi, the loss of forests, and the threats introduced by exotic species such as Australian trees, the Kurumba honey-hunting chants evoke a wistfulness for a time when honey hunting brought everyone together: "Oh my dear woman, mother, sister and everybody else/ Come and join in the feast / Let us sing and dance / We have collected honey successfully."[29] The chants become much more important in the face of such threats, not as a means of escape but as an affirmation that the Kurumba respect and celebrate their cultural heritage.

Honey collections are rarer than ever before. In his book *Honey and Dust: Travels in Search of Sweetness,* Piers Moore Ede documents

Apitherapy

Although not widely endorsed in the United States, apitherapy is a field that covers a wide assortment of bee-related products or treatments around the world. Perhaps the most famous beekeeper to popularize the medicinal use of bee stings was Charles Mraz, who began to offer apitherapy courses in the United States during the 1930s. The newest generation of apitherapists promote alternatives to conventional medical treatments for ailments such as arthritis and multiple sclerosis. Reyah Carlson travels to US bee conferences, but India offered new challenges:

> I arrived at the humble and lovely home of what is now considered my new family. I was greeted by many family members, some of them having driven 400 miles to meet me. I was blessed by "Grand-mama," the matriarch of the family who lives down stairs of the house which is part of what are known as societies, I believe we would refer to [them] as neighborhoods. She "blessed" me with incense, lit a candle, and then placed a red spot on my forehead, and said to me . . . "namaste."
>
> After being introduced to each family member, about 15 of them, I finally met the reason of my trip. Dipti Waghela was lying on her back in a bed due to the crippling affects of MS, tremors in her head and right arm, the all too familiar face of this horrible disease. . . . All of the family and friends had been there to witness Dipti receiving her first sting. The drama of the whole thing almost seemed like a let down when nothing happened. . . . The next day we were able to locate a beekeeper about 100 miles away in the country. . . . The beekeeper put about 200 bees in the cage I brought with me to sustain the bees. The next day, and every day while I was there, Dipti received stings, her husband now comfortable applying them. I also let him practice on me, as I can tolerate many stings with no problem. I wanted him to be comfortable with what he was doing.
>
> We were able to locate a person in a village much closer, about 25 miles driving distance, who doesn't mind robbing the wild bees [*Apis cerana*]. We took the cage to him, and he scooped up about 800 of these miniature versions of our bees, into the cage which maintains them quite well.
>
> Dipti's progress was heart warming. She was able to sit up, the tremor obviously lessened, and with assistance, able to walk out into the living room, thus, finally breaking out of the hell she had been trapped in for the last four months. ("My Trip to India")

the difficulty in arranging such a honey hunt and the extreme danger that it entails.[30] With deforestation an ever-present threat, India also continues to suffer massive famines and increased debt because genetically modified crops changed the agricultural paradigm in the late twentieth century. In an effort to address obvious agricultural shortfalls, Vandana Shiva, a physicist and food rights advocate in India, pleads for a call to action in which democracy is redefined in terms of agriculture. In *Stolen Harvest: The Hijacking of the Global Food Supply,* she implores: "A food democracy that is inclusive is the highest form of equity and democracy. Such a democracy can feed us abundantly because other species do not feed themselves at our cost; they feed us while they feed themselves."[31]

Such a food democracy, if feasible, may look like the industrial agriculture of the United States. But if so, it will have to depend on honey bees, and as the United States shows, a food democracy does not mean that women experience more equity. Much more extension work like that of the Darwin Initiative and Winrock should be made available to women in India so that they have more control of their local resources. India used to be known as the "heart of the world," and for good reason.[32] In just one region, the Blue Hills (or Western Ghats), four thousand species of flowering plants are known.[33] Of those, fifteen hundred species are endemic. Other authors disagree, claiming either that there are more endemic species or that there are fewer. The Keystone Foundation offers the following assessment: "Although the exact number keeps varying with the author and time, what is of interest is that nearly 38% of all species of flowering plants in the Western Ghats are endemic."[34] These floral resources will be the true keys to women's success in beekeeping, and local initiatives to protect the flowers will also inadvertently provide opportunities for women if education is accessible.

Asia

A Peaceful Renaissance

Most people don't even realize what we have right here on our own doorstep.
—Fahriye Hamulu, in *The Honey Spinner,* 2008

Although much about women and beekeeping remains shrouded in mystery, we know Asia was the first continent to correlate honey bees with goddesses and icons. Since Asia is the largest land mass in the world, it had diverse theologies that created divinities such as the ancient bee goddesses Artemis and Hannahanna in Asia Minor and Hitam Manis in South Asia. Much later, these goddesses would make way for three major monotheistic religions—Judaism, Christianity, and Islam—that would flower in the Middle East and spread their ideas about bees and women around the world. Given the diverse theologies in Asia, author T. R. Reid states in his book *Confucius Lives Next Door,* "the peoples of ancient Asia, i.e., Chinese, Japanese, Indians, Persians, Arabs etc., never conceived the idea of Asia, simply because they did not see themselves collectively. In their perspective, they were vastly varied civilizations, contrary to ancient European belief."[1]

While acknowledging ancient theological bee icons, in this chapter I skim the surface of these theological and geographic complexities by focusing on contemporary women and beekeeping. Only recently have women begun to appear with the force and autonomy that the goddesses expressed in ancient times. Beginning with Southeast Asian countries such as Malaysia and the Philippines, which are in the midst of an East Asian renaissance, the chapter moves to North Asia, where Russian beekeeping influenced much of central Asia in the late nineteenth and twentieth centuries as well as current beekeeping efforts in the region. The chapter concludes with Turkey, home of the Great Mother Artemis and gateway to Europe. Although I do not want to oversimplify, this chapter stresses that contemporary women beekeepers face theological, political, and economic challenges squarely at odds with the ancient agricultural feminine icons that once pervaded the regions.

The reason any generalization about Asian women beekeepers is so fraught with potential pitfalls is that, quite simply, there is still much to be learned about the bees in Asia's distinctive regions. Just as the Greek historian Herodotus shortchanged Asian people by lumping Asia into one word and one land mass, Asian honey bees have been shortchanged too. There is amazing diversity in Asian honey bees. Most people never realize that there are twenty thousand kinds of bees in the world. Six hundred of these make honey. Even as recently as 1988, most scholars assumed that there were four species of honey bees in one genus. Now entomologists acknowledge nine species, four of which have been found in Asia.[2] This diversity of honey bee species has affected the success or failure rates of international efforts to introduce beekeeping in Asia. Although most international agricultural development programs seem intent on introducing *Apis mellifera mellifera* (German black or European honey bee) or *Apis mellifera ligustica* (Italian), which evolved in Europe, these subspecies do not withstand Asian pathogens very well. More recent efforts acknowledge *Apis mellifera cerana*. And although information about Asian women beekeepers and their bees is not as accessible as that on other continents, there is no doubt that an Asian renaissance is happening.

Southeast Asia

The legend of the goddess Hitam Manis is not just a cautionary tale of love. It is also, depending on the teller and the audience, a teaching tool used to transcend social differences or pull generations together to learn ancient honey-hunting methods.

In this tale, a prince falls in love with a servant girl. In an effort to dissuade the young prince, his father turns the girl into a honey bee. The father's intervention is thwarted when the prince finds the girl-turned-honey-bee and takes her home to care for her. The twist in this tale is the unexpected sisterly relationship between the prince's wife, Fatima, and the honey bee Hitam Manis. The power of this ancient tale is still potent, it seems. Scientific scholars Oldroyd and Wongsiri blend the myth into their book *Asian Honey Bees*. Paraphrasing their description, the tale goes as follows:

Eventually, the prince married a real princess named Fatima, who liked her husband's bee and played with it among the flowers. Fatima took the bee to a tall tree that would be its new home. Here, no harm could come to it.

Fatima then decided to collect honey from the tree to use as medicine. So, together with the prince, she received instruction from the black forest and white heavenly spirit worlds on how to climb the tree and collect the honey. The prince proceeded to climb the tree with a bucket and rope. However, when the bucket was lowered, Fatima found not honey, as she had been expecting, but the prince hacked into many pieces.

Fatima then called to the tree spirit, "O spirit of the Tualang tree, restore my young shaman to life or I will utter a curse and there will be no more Tualang trees in the world."

The spirit replied with a deal: "I will restore the young shaman to life, but you must promise that from now on no knives or iron implements of any sort will be used when climbing a Tualang tree."[3]

Oldroyd and Wongsiri include the tale to show its impact on Asian honey-hunting cultures, explaining the origin of the bees, the need to placate the bees, the proper care of trees, and the constant interchange between spirit world and human world. However, its emphasis is primarily on Fatima, who will be responsible for reinforcing the rule of proper tree and bee management.

The beautifully illustrated children's book *The Bee Tree* by Stephen Buchmann and Diana Cohn tells of the beauty that can come from violence and elitism. In this retelling, Hitam Manis still suffers from the same social pressures related to falling in love with a sultan in a higher social class. To protect the royal class lines, soldiers are sent to kill Hitam Manis, and so she flees. Attempting to avert potential and certain violence, Hitam Manis metamorphoses into a bee swarm, destined to live outside the confines of society and high in the trees for protection from soldier's knives.

In this book, Hitam is called the "Dark Sweet One," and her power shines through in the book's pointillist illustrations. On one page, her image is "watermarked" behind a group telling stories around a campfire, and her tears fall onto her prince on another page. These superimposed images suggest the shadows in which many of our beliefs are formed and layered through the years.

The story of Hitam Manis in this retelling is encompassed within another story, that of a grandfather teaching his grandson Nizam the honey-hunting traditions. Fatima is not a character in the tale. Updating the story for contemporary audiences, Buchmann and Cohn place more emphasis on the honey bees, the proper way to harvest honey, and the importance of all people's respect for the trees. This tale has one clear message, a Malaysian proverb: "You cannot love what you don't know."[4]

Hitam Manis resonates in Asian literature for good reasons. For a region that has known much deforestation and foreign rule, her character emphasizes ecological respect, interrelationships with the environment, and a belief in one's actions enduring long after a temporary life. These versions stress that women, especially, have a responsibility to ensure the protection of trees and forests, for in their feminine spirit, the honey bee thrives and provides. Even Fatima, the prince's bride, learns that her former competitor, the honey bee, needs diligent protection.

These myths may seem out of place in contemporary Asia. Ching Crozier, who grew up in twentieth-century Malaysia, is neither servant girl nor metamorphosed goddess. She is not even a beekeeper in the typical definition of the word: she works in a laboratory at James Cook University in Australia. Yet, her research in the 1990s helped map the mitochondrial section of the honey bee DNA. Because this section contains matrilineal codes, it is the powerhouse strain of the DNA. Her husband, Ross, was a theoretician who used her research to make links with prior genetic research. Together, they established one of the stepping stones to mapping the honey bee genome sequence.

When asked about her background, Ching Crozier explained that becoming a scientist was a goal for Malaysian students regardless of gender: "In our days, it was quite common among Asian students to want to go into science based disciplines. At high school, students saw the East as backward because of a lack of understanding of science and technology and most high school students wanted to do science. My family supported my decision. I was encouraged to go to university."[5]

Although it was common for many male students to leave Malaysia to pursue advanced studies in the 1960s, it was more unusual for women to leave. However, Ching decided to go to Australia, where

Mitochondrial Genomes: The Powerhouse of the DNA

Oldroyd and Wongsiri explain the Croziers' research this way: "Mitochondrial genomes are circular DNA molecules, which in honey bees are about 17,000 base pairs long. . . . There are multiple mitochondria in every cell, which makes mitochondrial DNA extraction and preservation less difficult than it is for nuclear DNA, of which there is only one copy in the nucleus. Mitochondrial genomes are particularly suited to the examination of recent evolutionary events because they evolve quickly" (*Asian Honey Bees*, 59).

she met Ross at Melbourne University. When Ross returned to the United States to study at Cornell University, they continued to correspond. He eventually asked Ching to marry him, and she came to the States.

Following her husband's academic career, Ching worked as an assistant at Cornell and then as a technician with biochemist Jim Travis at the University of Georgia. Ross and Ching had two children, and in 1975, they moved back to Australia to join the University of New South Wales in Sydney. Ross described the challenges Ching faced in the 1970s and 1980s:

> Initially I hoped to sequence an mtDNA but we switched to honeybees. This was in the pre-PC era, and before that of automated sequence, so Ching had to clone and then sequence all of the mtDNA. It was heavy going, but a couple of significant papers came from that period. Then in 1990 we moved to La Trobe University and the work picked up speed. We published the results of Ching's labors, the complete mtDNA sequence in 1993. . . . Interestingly, it was Ching who spotted the possibility for a directly applied paper arising from the mtDNA work, and this appeared in 1991. I think it is the most directly applied of all my papers! It's on how to detect Africanized bees using mtDNA.[6]

Although Ching may not fit conventional images of a beekeeper, she deserves the title "honorary beekeeper" for her research with honey bee mitochondrial DNA. The honey bee is just as threatened and in need of protection now as in the Hitam Manis myths, and Ching Crozier's laboratory work has been a huge step in providing the environment to achieve that.

DIANA SAMMATARO: THE PHILIPPINES AND THE PEACE CORPS

To the east of Malaysia, the Philippines has been as active in participating in the Asian renaissance as its neighbors. When international agriculture programs were planned in the 1960s and 1970s to provide leadership and development, the Philippines invited beekeepers to provide extension and economic development classes. One of these was the North American Diana Sammataro. At that time, the Peace Corps unit was trying to bring in European honey bees, even though Asian honey bees *Apis cerana* and *Apis dorsata* were the dominant species. Since Asian honey bees already had varroa mites, Sammataro could see that the European bees would not do well at all. "It was so bad," stresses Sammataro, "the varroa mites were literally tearing the *Apis mellifera* bees apart." Furthermore, a varroa mite treatment Sammataro was trying from Hungary was completely useless. She could see the future for US beekeepers and concluded, "It was not going to be pretty."[7]

Other bee-related complications occurred. Another type of pathogen, *Tropilaelaps clareae,* was also affecting *Apis mellifera* bees that the Peace Corps had sent. There were not enough European honey bee colonies to set up mating yards, so between the low numbers and the mites, Sammataro experienced much frustration in basic bee maintenance.

Added to those stresses were cultural differences. For much of the recent past, the Philippine people have grown sugar cane for their dietary needs. Honey was considered a medicinal substance and used much more rarely than sugar. It was difficult to teach the Philippine people the dietary benefits of honey and the potential uses of beeswax when they had no beekeeping tradition.

A related challenge was teaching the use of conventional hive boxes, when for much of the local honey needs, the people hunted for bee trees. "The Filipino people did not want box equipment. It just was not what they were accustomed to doing," Sammataro explains.

Another societal challenge at the time was Project Green. This movement encouraged Filipino farmers to use chemicals such as fertilizers and pesticides, some of which were being discarded from more developed agricultural countries, primarily the United States. The idea was that if third-world farmers had access to chemicals, their production yields would go up. However, there was no training for the farmers, so they were applying chemicals on orchards indiscriminately.

Still, Sammataro had successes. She struck a deal with a mango farmer: she would bring her bees to his farm if he would go one season without using chemicals. To his astonishment, every single mango tree produced fruit.

Sammataro returned to the United States in 1979, but she still continues to benefit from her work in the Peace Corps. The people she befriended in the Philippines correspond with her fairly regularly, and her time there prepared her for the arrival of varroa mites in the United States. (I discuss this aspect of Sammataro's research in "North America: The Great Experiment, Part 2.")

LILIA DE GUZMAN: THE PHILIPPINES

Lilia de Guzman became fascinated with honey bees in the Philippines around the same time that Diana Sammataro was finishing her Peace Corps assignment there. De Guzman currently is a team member of the Russian honey bee project, a North American initiative intended to bring more diversity to honey bee populations in the United States. But her journey to Baton Rouge, Louisiana, began in her native country of the Philippines in the early 1980s: "I began my fascination with honey bees about 30 years ago when I was a junior in college. I earned a Bachelor of Science in Agriculture, Major in Crop Protection [Entomology], in the Philippines. The university had an apiary with about three *Apis mellifera mellifera* and I think five colonies of *Apis mellifera cerana,* our native bees."[8]

Key to shaping de Guzman's career was the university beekeeper, not necessarily the honey bees. "The beekeeper was so nice," de Guzman writes, "patiently telling me everything about the biology and importance of honey bees. The *Apis mellifera mellifera* was very gentle; we examined colonies without veils! I was so surprised and fascinated! I never thought they can be as nice as that because growing up in the farm we used to throw stones at feral native bees and then ran as fast as we could!"

Although de Guzman was not able to do a research project on honey bees during her college years for lack of hives, she was hired

Lilia de Guzman, bee researcher, works with the Russian bee team at the USDA Baton Rouge bee laboratory. Courtesy of Amanda Frake, photographer.

after graduation in 1981 to be a research assistant for an ecology professor and an acarologist. "That was then my introduction to the field of mites! Unfortunately, we were not able to conduct bee and mite research because of the lack of financial support," she writes. De Guzman experienced the same type of frustration that Sammataro experienced in terms of not being able to control varroa mites on *Apis mellifera mellifera*.

De Guzman brings skills to the North American varroa mite

control discussion that stem from her childhood days in the Philippines: "Coming from a developing country and born to poor farmers, I worked in the farm before and after school, rain or shine, and also was obligated to do chores at home at a very young age. I did all of my chores without complaints because we were not supposed to complain."

She relates: "Also, I did not have the basic things, including books and toys that Americans take for granted. A simple project seemed to take forever because of the lack of money to buy tools or materials necessary to accomplish the job. I learned to be creative, and with patience and hard work oftentimes work gets done. . . . Time is precious for me. I guess I got it from my parents, too. They always told us that as long as we finish the work then we can play as much as we want later."

"I have three older sisters, one older brother and one younger sister," de Guzman says. "My parents were small-time farmers; we didn't own acreage of land. During the rainy season, we planted rice, but during the dry season, we planted all sorts of vegetables—tomatoes, eggplants, string beans, mung beans, several varieties of gourds and corn. For years we also planted tobacco, but it required a lot of work so we convinced our father not to plant tobacco again."

Unlike in the United States, where so much agriculture is mechanized and chemically controlled, farming in the Philippines was done manually. "I remember waking up very early to either water plants or harvest vegetables for the market before going to school," explains de Guzman. "If we did not finish, we continued after school until dark. We used to use water buffalo in the farm, but later as my father become older, one of my sisters and I bought a tractor to help them on the farm. My parents are now gone, so my sisters and their spouses took over."

Because of her experiences on the farm, de Guzman wanted to choose another career, something that did not require intense manual labor: "I promised myself not to work on the farm for a living like my parents did. I was so sick of being very poor! I noticed that the 'well-to-do' people in our town were in the medical field. I envied nurses who were paid high salaries and had a chance to go abroad. I also thought that wearing a white uniform and a cap was just so awesome and looked very professional! So I wanted to be like them, hoping that I could help my parents get out of poverty."

But de Guzman's dream of being a nurse was not financially feasible: "My mother cried because there was no way I could fulfill my dreams. Fortunately, my sister Ely was willing to pay my way to college in addition to paying all of my brother's college expenses. It was definitely a sacrifice for her, so I opted for the cheapest way to get a four-year degree—go to an agricultural university."

Ironically, de Guzman's dream of wearing a white uniform did come true. "Here I am now—a bee researcher!" she remarks. "I still use a white uniform although it is a bee suit! Instead of a cap, I wear a bee veil! Instead of medical gloves, I have bee gloves. Rather than a scalpel, I use a hive tool. I certainly don't look professional this way but I am very happy with what I do."

De Guzman is so determined to succeed that she has been known to bring home a microscope to dissect bees for trachea mite analysis. "There was one occasion," she remembers, "New Year's Day, when we did not have any prepared food. I felt really bad when my children came to me and asked what we were having. I did not have time to cook; I just did not want to stop with my dissections. But I did make up for it at a later date. Yes, I dissected thousands of bees in my dining room. So, when the twins were in pre-school, the teacher asked, 'What does your mom do for a living?' The twins replied, 'She kills bees!'"

Even though de Guzman is a respected scientist in North America, she maintains a steady faith, another characteristic that she credits to her native culture: "I was raised to believe in asking for spiritual guidance in whatever I do including work. It may sound weird, funny or ridiculous to a lot of people, but I must admit that prayer gives me hope, strength as well as a sense of protection. It's a very powerful stress-reliever for me, very gratifying. This is some-thing that works for me."

De Guzman attended graduate school at Oregon State University in 1986 and at Louisiana State University in 1989. She has developed her career in North America working with genetic resistance to varroa mites. (For more about de Guzman's experiences as a honey bee scientist, see "North America: The Great Experiment, Part 2.")

Northeast Asia

The North Asian countries brought more organization to forest-based beekeeping than those of the southeast about three thousand years ago. Yet even with centuries of forest-based beekeeping records such as wax inventories from churches and bills of sale, Russian women were not recorded in early bee myths or ceremonial songs. Only when Catherine the Great came to power in the eighteenth century was there a Russian feminine advocate for honey bees and beekeepers. Her influence lasted, albeit anemically, through the nineteenth century. Twentieth-century Soviet communist policies changed that trend. *How* the Soviets changed the trend is worthy of consideration, even though the centralized framework no longer exists.

Enchanted with the latest advances in European beekeeping, Catherine the Great introduced a number of proactive measures to protect bees and beekeepers. Her first act was to protect the Russian forests, home to thousands of colonies of bees. According to European writer Dorothy Galton, "In 1740, she issued an *ukaze* declaring that lands where trees with hives grew were not to be allowed to become private property."[9] Nor was her interest in beekeepers a passing whimsy. In 1753 she ended all internal customs duties, including those on honey and wax. Catherine the Great followed her generosity with a manifesto in 1772 in which beekeepers were freed from all taxes: "Where taxes are still collected on tree beekeeping and apiaries, We remove them and order henceforth that there shall be no collection or payment." Apparently, Catherine was so fond of bees that "when founding the Free Economic Society in 1775 she gave it a coat-of-arms . . . that contained a beehive."[10] Under her reign, the first works of Russian beekeeping were written by Peter Rychkov. Previously, the only beekeeping material appearing in Russia had been translations of texts published in other countries. Catherine the Great's efforts provided a much-needed political foundation for the beekeeping arts to survive until the nineteenth century.

Russian beekeeping stalled during the nineteenth century, although there were pockets of beekeeping education and authors concerned that beekeeping be taught to everyone, including women. From 1800 to the 1850s, Nicholas Vitvitsky, a philosophy professor, taught beekeeping at the Lisinski Forestry School. In 1845 he published *The Glass Hive, or a Selection of Curiosities from the Natural History of Bees.* In describing the book, author Naum Ioyrish emphasizes it was aimed at all ages, all estates, and both sexes. In 1849 Vitvitsky took over four thousand colonies from a retired beekeeper. Like Catherine, he had traveled widely in Europe and wanted to educate all people, including women.[11]

One of the most famous Russian writers, Leo Tolstoy, married Sophia Andreevna in 1862, but he had fallen in love with honey bees much earlier in his life. In Sophia's diary in 1865, she lamented a type of "widowdom" that happens when a woman falls in love with a male beekeeper: "The apiary has become the centre of the world for him now, and everybody has to be interested exclusively with bees."[12] Still, she faithfully served as secretary, proofreader, and financial manager while he wrote *Anna Karenina* and *War and Peace,* both of which involve beekeepers or bee metaphors.

In terms of accessibility to apiculture, Sophia Andreevna was an exception. Russian women, especially peasant women, did not have many opportunities to develop as beekeepers in the nineteenth century. Serfdom was not abolished until 1861, and even then, women did not receive equity in pay, benefits, and other basic services. Bee schools were not widespread, although Peter I had developed extensive bee gardens in Ismailovo Apiary. The Russian Beekeeping Society did not form until 1891. The first beekeeping institute, Boyarka Technical School, was not established until 1902 in Kiev, under the leadership of Alexi Andriyashev. So there was no widespread network in which Russian women could be formally educated in beekeeping before the twentieth century.

The twentieth century brought major changes for women in the Russian Empire. Beginning with the revolution of 1905—that is, the creation of the State Duma—Nicholas II allowed for the formation of political parties, civil rights for all, and universal suffrage.

The revolution of 1917 replaced the czarist regime with the collective Soviet Union, and despite the atrocities and loss of life, in some ways women benefited from the transition. On International Women's Day, March 8, many women textile workers who were already protesting the czarist conditions began the first stage of the revolution. The main catalyst for the protest was the lack of bread; however, other grievances were defined as many women textile workers called for others to join in their protest. The march to socialism was on; women's freedoms continued to the beeyards and bee laboratories with more leadership than ever in the history of Russian beekeeping.

Lenin was determined to provide electricity to the Soviet Union in an effort to unify the disparate pockets and distribute information throughout the Soviet states. He also was determined to provide a new type of economy in which agricultural workers could earn more money by selling produce to the state. The New Economic Policy of the 1920s was considered one step away from capitalism, but it was more successful than the feudal estates had been. Rather than taking all of the agricultural goods workers produced, the state cooperative would take a percentage, and the farmers could sell the rest. Women went to school, learned to read, and participated in the Soviet Union's emphasis on research and agriculture. Industrialization provided some women economic equality, and women enjoyed more political and legal equality than before. Women had access to hospitals in which to give birth, day care, and other amenities that had been unheard of before the twentieth century. Literacy and worker-training campaigns meant that Russian women had better health care and economic options.

Socialist writer N. M. Glushkov notes, "In 1919, soon after the establishment of Soviet power, a special decree was issued by the Government of the 'preservation of beekeeping.'"[13] Russian women contributed to the scientific body of knowledge. From 1910 to 1925, Vsevolod Shimanovsky's wife assisted with the bee farm duties because Shimanovsky had gone blind. That she remains unnamed is surprising, but Shimanovsky was one of the teachers in the Boyarka School during the Russian winters, starting with one of the first twentieth-century research stations in 1919.[14]

Another woman, L. I. Perepolova, published her findings regarding trachea mites and Russian bees in 1929 and continued that research until the 1930s. Together with her husband, F. A. Tyunin, and two other staff, Perepolova directed the research with fifty hives of bees but very little equipment. In 1920 they moved to a new laboratory that was built on a farm near Lekhnikevich. Their research was mostly on bee forage, but the research conditions were rather spartan since they worked in a four-room building, two rooms of which were labs. They had oil lamps but "no piped water supply. . . . Eventually electricity was supplied and they contrived to fix up a water supply themselves as well as contriving to make most of their scientific equipment."[15]

When Joseph Stalin ended the New Economic Policy in favor of collectivization in 1929, Russian beekeeping suffered. There was no central research unit collecting the work of the bee stations until 1931, when the All-Union Lenin Academy of Agriculture was set up. The Tula and Moscow stations formed the basis of the Beekeeping Research Unit. Although some of Perepolova's research with swarming was referenced in *Bee World* in 1947, such references are few and far between.

Writing for *Bee World* in 1959, Glushkov details the "factory model" applied to beekeeping: "Queen breeding and bee-rearing nurseries are concentrated in the south of the country . . . hives are produced in government factories, in workshops, and by beekeepers themselves. . . . There are four factories producing beekeeping equipment and ten wax factories and workshops engaged in processing raw wax and in making comb foundation."[16] In describing how the factory model applied to honey production, Glushkov explains, "The main bulk of honey sold is extracted, either clear or granulated. The greater part of the honey obtained on collective farm apiaries is divided amongst the members of the agricultural association, and the surplus is sold to organizations which store the honey for future sale. The selling price of honey in the Gvt. Trade network is fixed according to different zones of the country."[17]

Added to the pressures of collectivization were the pressures of World War II. According to Glushkov, "There were 10 million colonies in the Soviet Union by 1940. WWII reduced this number by half when the Germans overran the central and southern regions of Russia. By the 1950s, there were more than 9 million colonies, half of which are on state farms and collective farms, and the remainder with individual collective farmers, workers, and employees."[18]

At the time Glushkov wrote his articles, there were "five experimental stations and ten departments and laboratories of beekeeping directed by the Beekeeping Research Institute," as well as "free six-month non-resident courses for improving the knowledge of agricultural specialists" and fifty-seven one-year schools of beekeeping "to prepare those who are setting up apiaries."[19]

In 1971, Dorothy Galton documented the following example of beekeeping on a collective farm in the subforest zone of Kemerovo region, noting that wives were some of the bee crew: "The bee farm was set up in 1964 with a staff of five. . . . In 1968, the farm had 3100 stocks of bees distributed in 30 apiaries in the forest zone, in places where trees had been felled, 25–100 km from the central farmstead. In each apiary the apiarist has helpers for 120 days in summer; and many of these helpers are wives of beekeepers. Each apiary has horses for the beekeepers to ride."[20]

So much industrialization and so few individual faces. It is poignant that when Eva Crane recognized L. I. Perepolova at an Apimondia meeting, Crane lost her characteristic British reserve and experienced a rare moment of fellowship with Perepolova: "When we met we both dropped everything we were carrying and embraced each other."[21]

Just as the Soviet factory beekeeping model seemed to be in its prime, other forces were at work that would undermine this system, namely, deforestation of bee trees and other forage. According to forest historian Michael Williams, "By 1950 the Soviet Union was suffering a unique form of deforestation that was set to get much worse in the subsequent decades. It was not brought on by the pressure of peasants seeking land, the influx of unemployed urban dwellers into forested areas, or plantation agriculture, but by the unwieldy centralized planning apparatus, where the size and multiplicity of forest cutters caused destruction over vast areas while paralyzing production that never satisfied needs."[22]

On top of the destruction of forests, World War II devastated the beehives and collectives. Nevertheless, Soviet writer Iosif Khalifman wrote with nationalist pride and scorn for Western beekeeping ways in 1951: "Soviet people cannot regard the economics of beekeeping in capitalist countries as anything but a string of absurdities. . . . In the U.S.S.R., the beekeeping industry has been placed on a firm foundation for the first time in history; the apiary is a planned branch of collective farm and state-farm economy and beekeepers have at last been enabled to discard amateur practices and devote themselves heart and soul to beekeeping as a profession."[23]

To counter Khalifman's optimistic view (he was awarded the Stalin Prize in 1951), Galton quoted in the appendix to her book an anonymous author who wrote: "Equipment is difficult to get, its cost high and quality poor. . . . There is a lack of transport for migratory beekeeping, difficulty in getting permits for it, lack of marketing facilities."[24]

By the 1970s, Russia had formed distribution centers, with other countries serving as the places for production. Armenia, Turkmenistan, and Ukraine had legendary honeys and centuries of forest-based beekeeping networks. Some twentieth-century women beekeepers in these countries benefited from Russian state-controlled systems of honey distribution. Yet when the Soviet Union collapsed in 1991, an entire infrastructure in which women participated was also eroded. The following section on central Asia details some international attempts to assist in the establishment of women's bee cooperatives.

Central Asia

Throughout the twentieth century, Russia has shaped women's beekeeping in North and central Asia by forming distribution centers to market the wildflower honeys produced in the mountain regions. Russia's intermittent efforts to industrialize beekeeping and educate women drifted to the heavily Muslim central Asian countries. But this effort was hampered by the diversity of people in the Caucasus region. This area hosts so many nationalities that the ancient Greeks once advised merchants to bring three hundred interpreters to aid them in their dealings with the multitude of races.[25]

To read Alder Anderson's travel writing of 1907, it is little wonder so many people were drawn to the region. He describes an ideal spot in which microclimates abounded in close range: "While in the lowlands, the orange and lemon, and semi-tropical plants like tea and cotton, grow luxuriantly, at the higher altitude, shrubs and flowers are met with which could only be found elsewhere in the far north, and on the fringe of the snow line the flora of the tundras of Northern Russia is reproduced. The hillsides are clothed with tangled masses of the vine, here growing luxuriantly, untended by the hand of man while the common woodlands of less favoured countries are replaced by wild orchards of apple, pear, plum and apricot."[26]

One Russian writer, M. Semenoff, estimated there were more than forty distinct nationalities, including Cossack, Tartar, Circassian, and Armenian.[27] Two tribes in particular—the Abkhazians and the Ossetes—were renowned beekeepers. The Ossetes worshipped a bee goddess whose name had several variations: Meritta, Merissa, and Mereime. According to Hilda Ransome, she was "the mother of the Circassians' god of thunder." Another tribe, the Moksha Mord-vins, "paid tribute to Neskiper-ava, a bee-garden mother, as the pro-tective spirit of the bees."[28]

This Caucasian bee culture was primarily one of honey hunting, since the advances in movable-hive technology had not reached the region. "The wild bees of the Caucasus have learned wisdom in the persecution to which they have been subjected for centuries," writes Anderson, "and they usually establish their strongholds in the most inaccessible places, hollows in lofty forest trees, or openings in rocky crags which the hardy mountaineer can only reach when suspended by rope over a yawning precipice."[29] Before Russian nationaliza-tion, the tribes had defined a code of the forest in which bee hunters claimed rights to a community of wild bees. Rarely did the Russian state interfere with that code.

Nor did the Russian state interfere with the status of women. Anderson hints at the repression of women in his article regarding honey hunting: "The manner of gathering honey is primitive in the extreme. . . . The comb, honey, and dead bees are all raked out togeth-er by means of a specially made iron hook fixed upon a cane handle. The whole mass, called *bochechnoi,* is then placed in pails, brought for the purpose by women and girls, for the Caucasian mountaineer is far too grand a personage to demean himself by carrying a burden." However, Anderson records that women also kept bees: "Enormous numbers of tame bees are also kept in the villages, and tending them has been the special avocation of the women."[30] When peace came among the tribes, more men began keeping bees, although Anderson does not specify whether the paradigm shift meant women were displaced from this avocation in the process. In the photographs of Abkhazian peasants accompanying Anderson's article, four heavily veiled women sit beside their family outside the home.

The Russian Imperial Agricultural Society tried to introduce conventional movable-frame hives, but with mixed results. A primary fear of the peasants, it seems, was that the new methods could be used to better determine honey production and calculate new taxes against them. Anderson diplomatically explains the dilemma caused by such new technology in such a remote region: "A large proportion of the peasantry evinces the greatest reluctance to abandon its old world methods. Hollow gourds and *bezdonkas* are persistently used in preference to the modern hives recommended by the emissaries of the Agricultural Society. . . . The unfortunate instructors sent out by the Agricultural Society were severely maltreated and the model hives were destroyed by the wild Abkhases."[31] In an accompanying picture, a heavily veiled woman is working a *bezdonka*.

Just as suspicion of government oversight was entrenched in the Caucasus mountains, suspicion of women's rights was equally embedded in the Muslim culture. In *The Stalin Era,* Anna Louise Strong writes: "The toughest fight of all for women's freedom was in Central Asia. Here, women were chattels, sold in early marriage and never thereafter seen in public without the hideous 'paranja,' a long black veil of woven horsehair which covered the entire face, hindering breathing and vision. Tradition gave husbands the right to kill wives for unveiling; the mullahs—Moslem priests—supported this by religion. Russian women brought the first message of freedom."[32]

So, before the collapse of the Soviet Union in 1991, women in the Russian confederate states participated in the well-established honey production system and Armenia and Turkmenistan produced wonderful wildflower honeys that the Soviet Union distributed to various points across Russia. By developing honey networks in other "colonies," the Soviet Union could meet its honey demands.

When the Soviet Union collapsed, however, the entire beekeeping infrastructure imploded, and it remains in transition. The trains have become unreliable. The road system is defunct. Many Armenian beekeepers were used to getting paid regularly; now, these beekeepers have had to learn entrepreneurial skills associated with marketing, such as labeling, bottle selection, and assessing market prices, and some specialized bee skills such as queen bee production.

Philanthropic organization Winrock International provides aid

An Armenian beekeeper named Anahit uses former Soviet Union hives for commercial honey production. Armenia has a plethora of wildflowers that make it an ideal place for beekeepers. Courtesy of Ed Levi on assignment for USAID.

in teaching these skills. Ed Levi, whose efforts with the Boté people were mentioned in the India chapter, has been instrumental in teaching basic marketing skills in Armenia. Women beekeepers have many challenges to negotiate, however. Honey in central Asia is sold on the black market because beekeepers do not want to pay taxes, just as Anderson noted in the early twentieth century. Transportation is beyond local women beekeepers' control. They do have access to wildflower forage, although much of their equipment is huge honey-production hives. Their equipment has lasted through the difficult economic transitions.

Turkmenistan has been similarly affected by the restructured Soviet Union. In 2001 Mike Embrey and Bill Lord, US extension specialists, worked with the Farmer-to-Farmer project in Turkmenistan assisting women beekeepers. Winrock financed their trip. Embrey and Lord are the most articulate about the repression that existed before the change in regime:

> Because of the repressive political climate in Turkmenistan, it had not been possible until the year 2000 to form a beekeepers association or co-operative. On previous trips, two attempts had been made to teach beekeepers rudimentary honey processing and sales techniques, but none of these efforts were sustained after the consultants left. Winrock was able to form a women's co-operative in the year 2000, and we thought that honey would be a good commodity for the co-operative to sell. The object of the co-operative is income generation for women, creating new jobs where none existed before, and incomes are limited to approximately $1 per day.[33]

Apparently the concepts of markets and distribution systems did not work in Turkmenistan, so Embrey and Lord wanted to try a cooperative model. As they envisioned it, the coop would serve as both honey processor and distributor. "Turkmen beekeepers store raw honey in 50 litre aluminum milk cans. The cotton honey crystallizes rapidly," they explain. "To liquefy the milk cans of solid honey, we helped the women's co-operative build a hot water bath to melt the contents of three cans at a time, and then constructed a double walled bottling tank to strain and bottle the honey, all out of locally available materials."[34]

Unlike previous attempts at women's economic development, the cooperative seems to have found measures of success for Turkmenistan women: "Our latest update tells us the co-operative has been able to get their honey into a large Turkish department store that has recently opened in Ashgabat. The honey is selling well, and the women are very pleased with the profits from the honey business and the opportunities the additional income opens up for their families. The beekeepers from whom the co-operative are buying honey are happy too, as they begin to see cash income for their hard earned honey."

Even in the Muslim countries Iraq and Iran, women have been more involved with honey and bees than mainstream media portrays. An article from the 1950s written by Khalel Jasim indicates that both men and women were keeping bees in Iraq, primarily in woven baskets and earthen jars, since movable-frame equipment was expensive and was not in use until 1938.[35] Although the state of beekeeping was rather primitive, Jasim states that approximately one hundred beekeepers were in the area around Baghdad and the bees suffered no diseases.

Contemporary wedding ceremonies in these countries continue to use honey. In her graphic novel *Persepolis 2*, Marjane Satrapi portrays her Iranian marriage. After meeting with the mullah, she and her betrothed follow the traditional rites of marriage: "Tradition required that a happily married woman rub two sugar loaves above our heads to pass on her joy and prosperity." The next scene shows them dipping their fingers in a bowl of honey. "Tradition also required that we plunge our fingers in honey and that we suck one another's fingers to begin our married life on a sweet note."[36]

Recent political changes in Iraq have meant greater freedom for women to learn about honey bees. When the United States entered Iraq in 2003, National Guard specialists engaged in broad nation-building programs involved with apiculture, focusing especially on women. In a recent beekeeping class, "the graduating class had twenty students, and around eighty percent of them were widows," according to Fatimah Muhammad, who is from Chicago. "It's a skill set they can teach others," she said. "They can take it to their communities and empower other women and help them earn an income off of it. . . . Part of the follow-up for the graduates will be getting the honey sold locally and in the neighboring Babil Province, which has a higher demand for honey and even houses a honey factory." [37]

Even in Afghanistan, US women serving with the National Guard work toward developing beekeeping communities in which women will once again be prominent leaders in their communities. A project in Kapisa was set up to supply women with beehives and

An Iraqi woman attending a beekeeping class in traditional dress.

teach them how to manage the hives and harvest the honey. Initially, twenty-five women were each supplied with four hives. Whitney Hughes wrote about the project in her article "How Sweet It Is."

Hughes notes that in April 2010, the Kentucky Agricultural Development Team, who were helping the Kapisa women with the project, met with leaders to talk about how to proceed during the peak honey season, spring.[38] The beekeeping project was begun by the Nebraska Agricultural Development Team in 2009, and the idea is for it to become self-sustaining. The women beekeepers are to maintain the hives for three years, splitting the hives to increase the numbers. At the end of three years, the Kapisa women expect to double the amount of hives and return the original four hives to the Director of Agriculture Irrigation and Livestock office. "The original four hives will then be redistributed to 25 other women, starting the process all over."

The Agricultural Development Team provides the hives, the bees, and the training. Since the Afghan women beekeepers can pro-

Jo Lisa Ashley, Kentucky, United States. "We should have standing armies of bees," Christian Konrad Sprengel once said (Hans-Ulrich, "Retrospect"). Ashley, the women's empowerment coordinator of the Kentucky National Guard Agricultural Development Team, displays a frame of honey in Afghanistan, 2009. Ashley taught beekeeping to Afghan women.

Major Bobbie Jo Mayes, Kentucky, United States, women's empowerment coordinator for the Kentucky National Guard Agricultural Development Team, 2010–2011, took this picture in Kapisa Province at the Director of Agriculture Irrigation and Livestock compound.

vide that training to other women, the project constantly replenishes the stock and the teachers. This model has worked well with other international groups, such as the Heifer Project.

In Afghanistan, according to Hughes, "one bee hive can produce up to six pounds of honey per year, and in local markets it is selling for 400 to 1,000 AFG (Afghani) per kilo, or about $6.60 per pound. So, one bee hive is worth about $39.60 per year." Sergeant Jo Lisa Ashley, the officer in charge of helping the women, "stressed the social significance that the project carries with it." Ashley explained: "The women are rebuilding the self confidence that was lost in the thirty years of war tearing their families and social status to pieces. With projects such as this the women are working their way back up via the country's agriculture foundation to show they too are instrumental in the reconstruction of a war-torn country."[39]

An Afghan business woman named Hassina Syed visited the United States in 2010 to learn more about business principles she could take back to Afghanistan. She remarked on the difficulties for women starting businesses in her country: "Nobody gives any loans for a woman's business to grow. There are some micro-credit loans available in Afghanistan. The maximum loan is $20,000, which is so small an amount to have a long-term sustainable business for a woman. . . . They need more money."[40] Addressing the main social factor that will have to change in Afghanistan, Syed commented: "It's a male-dominated society. Nobody takes us seriously. [But] the new generation who comes from foreign countries know women and men have to work together; otherwise the country will not develop."

Asia Minor

Turkey forms a natural land bridge between Asia and southern Europe. Although it used to be called Asia Minor, Turkey, also called Anatolia, is the home of some of the first gardeners, the first agriculturists, and the first devotees of powerful female goddesses associated with honey bees. Reflecting feminine experiences such as virginity, birth, and motherhood, Artemis was a goddess who was worshipped by many peoples who traveled this land bridge. She is the original Great Mother of the Hittite culture.

The Hittite culture was in power from 2800 BC to 700 BC, and their cuneiform tablets provide the oldest laws governing bee culture. The Hittite laws defining property and propriety show a culture in transition, of which honey bees were a crucial part. To discourage theft, city officials distinguished between swarms, which were considered transitory, and hives, which were considered property. If someone stole a swarm, he had to pay five shekels of silver, or approximately one sheep, and had to submit to being stung as punishment. A revised set of laws mandated that the thief pay six shekels of silver, with no punishment by stings.

The same sense of propriety extended to Hittite women. Historian Hilda Ransome explains: "It is at this point in our study of the history of the bee that the belief in her chastity begins to emerge, a belief which lasted through classical times, through the Middle Ages down to quite modern times."[41] Chastity in the Hittite culture applied to men as well as women. Male officials connected with the worship of Artemis were called "Essenes," which means "king bee," although there is no such thing as a male "king" in the hive. These officials were bound to observe chastity for one year.

But it was Artemis nurturing a number of bee-related images, in full-bosomed fashion, that made the lasting impression on her devotees. Reproduction in particular was reconciled in her balance of animals and ideology. It was not known that queen bees mate in the air, outside the hive. Thus, "the mere fact that no primitive, and even very few modern men, have ever seen bees in sexual intercourse was sufficient to arouse all sorts of speculation about their miraculous origin and about their corporal virtue," explains Austin Fife.[42]

Artemis worshippers also believed in renewal and regeneration, using honey bees to understand reproduction since it was assumed that they could be begotten from bulls. Although the myth of ox-born bees has no truth, people may have "confused the larvae of honey bees with the larvae of blowflies *Calliphora* spp., or [confused] adult honey bees and drone-flies *Eristalis tenax,* whose larvae feed on decaying organic material," explains Eva Crane.[43] Nonetheless, people believed that bees could form a hive in the bull carcass without overt sexual actions.

Thus, the concept of chastity, a societal convention regulating feminine behavior using honey bees as a model, transferred across the land bridge from Western Asia to Greece. The goddess Artemis was even adopted by Crete and the Palestinian region before the Hittite kingdom disbanded. Ransome's research clarifies the link between feminine icon Artemis, Hittite culture, and chastity: "When the Hittite confederation broke up between 1200 and 700 B.C., many other religious influences were at work, but the ancient worship of the Great-Mother persisted, and became very pronounced at Ephesus, which had been one of the Hittite harbors. Here she was known to the Greeks as Artemis, and later to the Romans as the great 'Diana of the Ephesians.' This is not the Greek Artemis, but the old Asiatic Great-Mother, as is clearly shown in her archaic cult statues with their many breasts, which are the symbol of fertility."[44]

Other Anatolian goddesses such as Hannahanna protected Asia Minor from famine and misfortune. According to Grace Pundyk in her book *The Honey Spinner,* "When the moody god of agriculture Telepinu would storm off in a huff over some petty disturbance, Hannahanna would dispatch her bee to sting his hands and feet, wipe his eyes and feet with wax and bring him back to put things right."[45]

But these magnanimous goddesses no longer dominate the Anatolian plain. After centuries of Islamic theocracies and patriarchal leadership, women in this region of Asia have been encouraged to focus their attention on worship and domestic skills rather than agricultural arts, although new opportunities for women beekeepers are arising. When Turkey prepared to be the first Muslim country to be included in the European Union, a lot of international groups and money arrived to provide training and education, including in beekeeping. That Turkey was unsuccessful in its bid to be admitted to the European Union has not dampened people's desire for honey. Fahriye Hamulu is a Turkish beekeeper who keeps fifteen hundred hives near Istanbul. Throughout Grace Pundyk's trek around Turkey in search of honeys, as described in her book, Hamulu served as a "virtual" tour guide, using cell phone and e-mail technology to stay in contact with Grace. At various stops in the Turkish countryside, Fahriye gave the bus driver directions, arranged for tours and television interviews, and in the midst of political protests, managed to see a future beyond the immediate cultural and political challenges.

Seeing the EU membership as an opportunity to participate in a global market, Hamulu had wanted to establish a product of pride: "If I can make a laboratory here in Istanbul that follows the right and proper way of testing honey and only selects that honey which is pure and free from pesticides and other chemicals, then I will be happy."[46]

Hamulu's desire for a high-quality product and the freedom to run a business on her own terms point toward a Turkey in transition. Hamulu married "long enough to see the landscape," as Pundyk says diplomatically, and just long enough to become pregnant. When her son was sixteen, Hamulu had him move to his own place to develop independence. She did not learn beekeeping from her father or grandfather. She learned by herself because "I know how valuable our bees and honey are. But most people don't even realize what we have right here on our own doorstep."[47]

Hamulu's insight that Turkey is a "sleeping giant" for honey production is true. Europe and other Western cultures have not known the value of the Asian honeys, its bees, or even the potential of Asian women to be beekeepers. Asia enjoys the greatest diversity of honey bees and the most land mass to provide forage, and yet it has the most obscurity concerning women and bees. An observation of Pundyk's about Hamulu may apply to more than just one woman in Turkey: "In many ways, [Hamulu] reflected that other side of Turkey, the side that says, 'We are who we are, and we don't need the EU to be something.'"[48]

Pundyk continued to define Hamulu as a woman pointing to the future but also harkening to an ancient past: "[Hamulu] is a woman, and a fiercely independent, self-made one, and in many ways I felt she linked in with the ancient goddess traditions of this land, before it was Christian or Islamic or, even, Turkey."

Placing her experience in Turkey in a broader context, Pundyk explains:

> I saw honey and bees in Turkey as reflecting what's happening for them as a nation, on the global and political stage and historically. On the west coast of the country, where EU development money has been around for a lot longer than it has in the east, the pine honey market, for example, which accounts for around 80 percent of honey produced in the country each year, is more developed, more organised. In eastern Turkey, where it is much poorer, more rugged and wild, and bordering unpopular countries such as Iran and Armenia, development dollars are only now just starting to reach this area.
>
> What I think is telling in this regard is the fact that of the 200,000 tonnes of honey imported into the European Union each year, Turkey's contribution is only minimal. In 2005, it was only 18,000 tonnes. Certainly, there are lots of practical reasons/answers why this is so, but the fact that Turkey is so close to Europe and has such potential for honey production and yet this hasn't been fully developed by the likes of the EU, I believe is a reflection of a much larger, historical story.[49]

To put Pundyk and Hamulu's comments in a broader perspective, Turkey's geographic position places it at the center of an increasingly complex dialogue in which it could lead the twenty-first century. North American economist Thomas Friedman makes an important distinction in his "Letter from Istanbul, Part Two":

> Americans have tended to look at Turkey as a bridge or a base—either a cultural bridge that connects the West and the Muslim world, or as our base (Incirlik Air Base) that serves as the main U.S. supply hub for the wars in Iraq and Afghanistan. Turks see themselves differently. "Turkey is not a bridge. It's a center," explained Muzaffer Senel, an international relations researcher at Istanbul Sehir University. Since the collapse of the Soviet Union, Turkey has become the center of its own economic space, stretching from southern Russia, all through

the Balkans, the Caucasus and Central Asia, and down through Iraq, Syria, Iran and the Middle East.[50]

As the twenty-first century continues, the "Asian renaissance" will continue to provide more insights into the complex transfers from ancient myths about bees to women's roles. The contemporary retellings of ancient myths of Artemis, Hannahanna, and Hitam Manis establish that Asia once had a more equitable culture. These myths continue to define femininity by transferring values ascribed to honey bees such as chastity, equanimity, and stoic serenity to feminine experiences. These values have survived even though the environments in which they flourished have faced serious threats, both ecological and political. Massive deforestation threatens honey-hunting cultures. The breakdown of the centralized Soviet Union has meant some areas of Asia have suffered major economic setbacks for women beekeepers. Cooperatives are forming in lieu of centralized distribution centers, but much work remains to be done. Compared to other continents, Asia has fewer examples of individual initiatives. Similar to the naturally occurring Asian mineral reserves suppressed under layers of sediment, powerful bee-related feminine icons remain as touchstones for the twenty-first century. With the advent of women's rights and global outreach, more Asian women will draw strength from these underlying cultural "pools" and find in bees a way to define a future for themselves and their loved ones.

Europe

A Bridge of Honey Bees

I will not account her any of my good Housewives that wanteth either Bees or skilfulnesse about them.
—William Lawson, *The Country Housewife's Garden,* 1623

Europe has the most consistent, readily transparent, and best-documented history of women beekeepers when compared to the areas discussed in previous chapters. The process of tracking the societal transfers of bee analogies has been easier. As economies, theocracies, and social policies ebbed and flowed, the bee-related values first ascribed to women by the Hittite culture flowed west to Greece and Europe and later coexisted with northern Europe's iconic bee goddesses.

The highly gendered principles that emphasized a feminine landscape capable of providing for people were rooted in male fantasies of a golden age. In ancient Greek writers' visions of alternative family units, femininity was defined by a quality that historian Herodotus called "women's communism," a trait perhaps influenced by the many examples of honey-bee hives in Greece. Plato wrote of an ultimate family unit in which "all these women . . . belong to all these men . . . so that no father knows his child, nor does any child know his father."[1] In myths of Amazonian women, in which men's power was significantly limited, women worked as one social unit.

If the ideal was complete civil harmony, ancient writers seemed to think, then civic order was fundamentally connected with feminine roles of equity, cooperation, fertility, and service. The hive was the best example of such communality. These analogies, often written in times of great political stress, took form in feminine religious icons and myths that assuaged anxieties in times of great strife.

Southern Europe was where the gendered analogies between women and bees began to be a literary trope. Because the continent has had such transparent bee cultures, women's active participation and leadership in apicultural communities have had fewer consequences through the centuries. Whether in theological, social, or educational realms, European women have participated in beekeeping, creating a strong foothold for knowledge-based apicultural economies. Based on this history, European women continue to establish new directions in honey bee research, international extension, and honey-production standards.

Southern Europe: Goddesses, Nymphs, Nuns, and Queen Rearers

GREECE

As I conclude the writing of this book, Greece has shaken the European Union, a union of countries that was supposed to provide peace and financial stability. Greece's prime minister has realized that his country's overextended social policies finally need to cease in the twenty-first century. The European Union was a dream that followed a twentieth-century nightmare of two world wars and the resulting economic consequences, not to mention the emotional trauma layered in increments on an entire continent.

The 2010 Greek protests seemed like mere temper tantrums until bank attendants were killed. Now the anxiety has spread to Spain, Portugal, and France. Germany, a country that has shown fiscal discipline since its emergence from the destruction of World War II, has vacillated between offering financial stability or withdrawing from the European Union. So contemporary Greece is a far cry from the country that positioned itself to dominate bee culture in the 450s BC, the golden age. Yet the stress and anxiety surrounding that culture, the ideals regarding governance, the division of labor and work ethic—these problems the ancient Greeks resolved by applying practices they observed from their neighboring cultures and their own practices drawn from their observations about honey bees.

The ancient Greeks recognized a fundamental fact of beekeeping: bees swarm. The swarming occurs when an older group of bees leaves an established hive with a mated queen to relieve stress in the hive and start a new one. Since swarms mean lost honey production (the remaining virgin queen must have time to mate), most contemporary beekeepers do not want their bees to swarm.

Although they did not fully understand swarming, the Greeks respected it by devoting goddesses to the occurrences and systematizing their worship around bee swarms. The Greek goddesses made the association between fertility and bees explicit. According to A. B. Cook in "The Bee in Greek Mythology," Greek ideology appropriated three important natural phenomena—swarms of bees that dwell in caves, swarms that dwell in hollow trees, and occasional swarms that have built their combs in the dried carcasses of cattle and wild beasts.[2]

Some biological facts about the hive must be clarified before going much further. If hives are not making honey, they are making queens. If there is too much honey in the hive and not enough room for the queen to lay eggs, the hive will become too crowded and the bees will shift their attention from honey production to rearing young queens. Once this shift has occurred, the older bees will swarm, and once started, this is almost impossible to reverse.

Swarming had to have happened often in ancient Greece. By the time of Pericles, which was around 429 BC, ancient Attica, with its area of about forty square miles, recorded twenty thousand hives of bees.[3] In fact, according to a fifth-century comment of Plato's in the *Critias*, so many trees had disappeared that the mountains could now "keep nothing but bees."[4]

The swarming instinct continues to confound contemporary beekeepers who want to manage such behavior. When a queen bee begins laying fewer eggs, the worker bees pick up the shift in the queen's pheromones, and they begin feeding royal jelly to eggs placed in queen cups. In effect, they are tending future queen bees. These will develop into queen cells.

The older mated queen meanwhile takes the older worker bees, which are ready to build a new hive, and leaves the hive in search of another home. In a swarm, the bees are less defensive because they have no home to protect. Meanwhile, in the hive, a virgin queen will emerge from a cell and destroy the other queen cells. The hive will then begin a new cycle.

These are the unadorned basics of a swarm. The experience of seeing a swarm is quite sublime: swarming is one of the most phenomenal forms of reproduction nature has to offer. The sight and sound of those bees moving en masse will stun me silent.

The Greeks appreciated this lesson regarding honey bees: sometimes the universe gives miracles for free. These gifts need appropriate respect, and far be it from the Greeks to deny women a role in that. Many goddesses were associated with honey: Demeter, Artemis, Ceres, Hecate, Persephone, Aphrodite, Selene, and the nymphs, especially the naiads. Perhaps this is why Greek women had not only their choice of goddesses they could worship but also positions in which they could serve as priestesses. Zeus was reputed to have been fed by the nymphs Amalthia and Melissa, so that his priestesses were called "melissae."

Put simply, says Austin Fife, "there was something miraculous about the generation of bees. It was not accomplished by copulation. And yet to the casual observer their manner of procreation was extremely successful—so successful that, beginning with a mere handful of bees in the early spring, in a few short weeks they are able to fill their hives to overflowing and send out from one to as many as ten swarms in a single summer."[5]

It was not always easy being a beekeeper in Greek mythology, but there were tangible rewards. In Demeter's cult, one of her initiates, also named Melissa, refused to tell the goddess's secrets. Melissa was torn to pieces, but Demeter arranged for bees to emerge from her body. Fife argues that this myth, along with the fact that bees nest in caves and tombs, factored into the public perception about bees being an incarnation of a soul.

Apollo's priestesses at Delphi were known as "bees"; the second temple, famed for its oracles, was built of feathers and beeswax to pay tribute to the birds and bees, commonly believed to be able to deliver prophecies—obviously a myth created in an attempt to explain the sources of the temple's oracular powers. Pindar says the priestess who answered the inquiries of those who came there to consult the oracle was known as the "Delphic Bee." Likewise, the priestesses of Demeter, Artemis, and other divinities were also called bees.[6]

The ancient maiden priestesses called the Thriae also dwelled at Delphi. The *Homeric Hymn to Hermes* says of them: "There are certain honey ones, sisters born—three virgins gifted with wings: their heads are besprinkled with white meal and they dwell under a ridge of Parnassus. . . . From their home they fly now here, now there, feeding on honeycomb and bringing all things to pass. And when they are inspired through eating yellow honey, they are willing to speak truth; but if they be deprived of the gods' sweet food, then they speak falsely, as they swarm in and out together."

The nymphs came to the assistance of the god of beekeeping, Aristaeus. In the first few years of his life, the naiads fed him and taught him the art of beekeeping. In a myth of unrequited love, Aristaeus became smitten with Eurydice, engaged to be married to Orpheus. When Aristaeus attempted to rape Eurydice, she ran away, and in her escape, stepped on a poisonous snake. Orpheus lamented the loss so much that the gods killed Aristaeus's bees as revenge. Only after appropriate penance did the naiads help him regain his hives.

The Greek women were not all good, and bees were used to ward off the charms of evil women. In Homer's *Odyssey*, the Sirens were such a threat that Odysseus had to melt "honey-sweet wax" to save his crew from their songs.

With absolutely no knowledge of the biology of the beehive, Hesiod, who may have lived about 800 BC, compared women to drones. Eva Crane and A. J. Graham describe his reference as the first analogy between people and bees, and it is not particularly flattering to women: "As when the bees in the roofed hives feed the drones, which conspire to do evil deeds. Every day the bees work eagerly all through the day till sundown and set the white combs, while the drones stay within the roof hives and gather into their bellies the foil of others. Just so high-thundering Zeus has made women to be an evil for mortal men."[7] Crane and Graham mince no words when they label Hesiod a "thorough-going misogynist."

Fortunately, in a poem titled "Females of the Species," the poet Semonides balanced out the negative perceptions of Hesiod: "In the beginning the god made the female mind separately. He made one woman from a vixen, one from a bitch, and others from other female animals, but all had many faults." When Semonides wrote about the woman who had been blessed with the mind of a honey bee, he rhapsodized about the consequences for a human family: "The man who gets her is fortunate, for on her alone blame does not settle. She causes his property to grow and increase, and she grows old with a husband whom she loves and who loves her, the mother of a handsome and reputable family. . . . Women like her are the best of most sensible whom Zeus bestows on men."[8]

The Greek writer Xenophon provided the first textbook for young wives, comparing the house to a hive. This was not a "how to keep bees" book, as later writers would offer women, but it does suggest that women were readers and taken seriously as an audience.

Of all the Greeks writing about bees, Xenophon has the distinction of correctly identifying a queen bee. In the *Oeconomicus* VII, Ischomachus tells Socrates how he taught his young wife to be like the queen bee in a hive. In a dialogue, the young wife requests of her husband some instruction on her duties. In replaying the scene to his friend Socrates, Ischomachus replies in good-natured manner with "things of no small account, I fancy, unless indeed, the tasks over which the queen bee in the hive presides are of no small moment."[9]

Continuing with the analogy that anthropomorphizes the queen bee's role, Ischomachus provides a list of the domestic duties that the queen bee and his wife share:

> She stays in the hive and does not suffer the bees to be idle; but those whose duty it is to work outside she sends forth to their work; and whatever each of them brings in, she knows and receives it, and keeps it till it is wanted. And when the time is come to use it, she portions out the just share to each. She likewise presides over the weaving of the combs in the hive, that they may be well and quickly woven, and cares for the brood of the little ones, that it be duly reared up. And when the young bees have been duly reared and are fit for work, she sends them forth to found a colony, with a leader to guide the young adventurers.[10]

Xenophon gets the analysis half right: the monarch of the hive is a queen, but the leadership abilities he attached to her were wrong. Queen bees do not have regal authority. Just why Xenophon's insight about queen bees was not shared by his colleagues remains a mystery. So the world would have to wait until 1586, when Spaniard Luis Méndez de Torres described a queen bee. Shortly afterward, in 1609, English writer Charles Butler dedicated a book to the "feminine monarchy," and the *idea* of a queen bee finally emerged as a scientific possibility.

Unfortunately, the sculptures of Greek goddesses belied the social status of Greek women, who were often not far above their slaves in social position. A Greek woman's duty was to oversee the servants. One writer complained that "although the literature gives exquisitely beautiful portraits of ideal womanhood, still the general tone betrays a deep contempt for woman."[11]

With no rights to own land, attend universities, agree to divorce, or even name or educate their children, Greek women remained under a patriarchal system for centuries until the nineteenth century, when they were allowed to enter universities for the first time. Even then, progress toward equity was intermittent. Just when women seemed to be on the path to ratifying their right to vote, political leader Ioannis Metaxas became dictator in 1936. His regime was followed by World War II and a series of civil conflicts. Women would not receive the right to vote until 1952.

None of these difficulties mattered to the beekeeper Penelope Papadopoulo, a very visible success of the equity movement in Greece. Becoming a well-known extension agent, she began her career by teaching beekeeping in Crete. After she encountered male opposition to her programs in Crete, Papadopoulo began to teach the wives, who had much more success with honey production. Papadopoulo's long career in beekeeping extension work included an illustrious career in Rhodesia, now Zimbabwe (see the chapter on Africa). She blended pragmatic beekeeping styles, some of which were based on ancient beekeeping technology, in countries that were struggling with high rates of poverty, poor transportation systems, and low levels of technology. Yet she also tried experiments with controlled queen mating in drone enclosures. Papadopoulo represented a new type of Greek approach to beekeeping, one that combined both ancient and modern methods.

Perhaps women who choose to be beekeepers will not face such difficulties in the future. According to Nicoletta Pantziara, "The 1980s can reasonably be characterized as a turning point for women in Greece. The fall of the military dictatorship in 1974, the rise of the Greek Socialist Movement, and the contributions made by the new First Lady, the American feminist Margaret Papandreou, paved the way for a number of institutional and legislative shifts."[12]

When Greece joined the European Union, it agreed to modify its laws to EU standards. In particular, Greece was required to ratify Article 119 of the Treaty of Rome and the United Nations Convention on the Elimination of All Forms of Discrimination against Women.

However, problems still remain, especially for rural women, "who face great difficulties when they attempt to enter the labor market as something other than farm workers," according to Pantziara. Technical schools, which often offer beekeeping programs, still reflect gender disparities: "In Greece, very few women are admitted into technical institutions, which are seen as a male domain; instead,

Angeliki Galineas, Exohori, Mani, Greece, hauls Langstroth honey boxes up to the mountain apiary. The apiary is situated on the side of the mountain, where she has her garden and which has no vehicular access. Courtesy of John Phipps, editor, *Beekeepers Quarterly.*

women choose schools with a humanistic direction. These educational choices obviously close off certain options for Greek women as they enter the job market."[13]

Apiculture, as well as the value-added industries associated with bees, may not be seen as a career option by many women, but honey is still a feature of bridal ceremonies in Greece. "It is customary," according to Marina Marchese, "for a bride to dip her finger into honey and make the sign of the cross before entering her new home. This gesture would bring her a sweet married life and good relationship with her mother-in-law."[14] This practice, along with new standards that eliminate the wedding dowry, an option to divorce, and some discussion around surnames, again reflects Greek women adapting ancient traditions to new transitions. Although Greece is revising many fundamental policies toward women, its legacy to women beekeepers has been to assimilate femininity with beekeeping, creating icons in which bee goddesses serve as a benign passive ideal.

ITALY

Although the Roman writers Varro (116–37 BC) and Pliny the Elder (AD 23–79) were just as misinformed about the queen bees as most Greek writers, Roman women goddesses closely resembled Greek bee goddesses as icons promoting reproduction, love, and oratorical skills. In fact, the small marble statue of Diana of Ephesia in Rome's Museo Capitolino has rows of bees and flowers carved into her, and as Grace Pundyk succinctly describes, "her multiple breasts resemble the plump brood found in hives."[15]

Roman armies quashed Greek independence in 146 BC, but many aspects of Greek culture, especially apicultural aspects regarding feminine ideals associated with bees, did not end at all. These ideals, and beehive technology, spread as the Romans capitalized on the extensive Greek networks. In terms of women's rights, however, Roman culture was not much more advanced than that of the Greeks. Women were expected to be child bearers primarily, and their secondary function was to educate the children. This was a major difference between Greek and Roman cultures. Since mothers were nurturers, it was assumed, they should also teach Roman culture. Augustus was so concerned about the decline in female children that he passed laws against celibacy and not marrying: the Julian Laws (18 BC) and the Papia-Poppaean Laws (AD 9). But the Roman culture still emphasized the importance of male lineage, and the laws seemed not to help the status of women. One early twentieth-century writer, who went by the pen initials E.J.P, summed up the status of Roman women as those who lived in "almost Oriental seclusion."[16] In *Women and Politics in Ancient Rome,* Richard Bauman says that "the public position of women was so unfavourable that it has even been doubted whether they were Roman citizens. . . . Those doubts were unfounded."[17]

With this status of women as a cultural backdrop, Italy became increasingly Christian, and the ancient associations between women and bees took on formalized and institutionalized roles. By AD 313 the Emperor Constantine had authorized Christianity as a lawful religion, and in AD 325 he convened the Council of Nicea.

The temples to Diana, Artemis, and Ceres went by the wayside. In their places were a new list of names from Hebrew and Christian cultures: the prophetic Deborah, the Egyptian convert Asenath, and the Virgin Mary were the new icons of Christian ideals of chastity and purity. Honey bees served as mediators.

Similar to Greek and Roman goddesses of prophecy and intuition, the Jewish prophetess Deborah connoted her wisdom as a judge and her poetic ability as a prophet. The Hebrew word Deborah means "bee," deriving from the word *debash,* "honey." Along with General Barak, Deborah inspired the children of Israel to win a decisive battle over Sisera and the armies of Canaan. Her song of victory and praise of God has remained as one of the masterpieces of earlier Old Testament poetic literature.

Although the religious convert Asenath is mentioned only once in the biblical chapter Genesis, the myth reflects marriage rituals in transition. Asenath, a beautiful Egyptian, was very proud until she happened to see Joseph, the eleventh son of Jacob and a former slave who had been traded by his brothers. Joseph worked his way out of slavery to become the most powerful man in Egypt next to the pharaoh. Asenath, the daughter of a well-connected official, had her choice among the well-placed men in her father's court. But Joseph caught her eye. Initially, she rejected the idea of marrying a Hebrew, but then she felt remorse, for she intuitively realized he was a man of God.[18]

According to the story, Joseph fell in love with Asenath but had concerns about their religious differences. He initially refused to kiss her, relenting only when Asenath converted to Judaism. In a crucial paradigm shift, Asenath converted to Christianity with Archangel Michael acting as an intermediary. Her conversion served a primer of prerequisite rituals such as fasting, humiliation, and prayers. But the most important ritual, that of communion, is the one most clearly associated with honeycomb and its importance as a religious food in Christianity.

Compared to Deborah, Asenath has been a marginalized figure by scholarly standards. Some of this marginalization may have to do with the decline that organized beekeeping was going through and a parallel decline in the importance of honeycomb in Christian rituals

Honeycomb and Communion

Before the marriage between Joseph and Asenath was to take place, Michael instructed Asenath, "Enter thy storehouse and thou wilt find a bee's comb lying upon the table; take it up and bring it hither." The story continues:

Asenath entered her storehouse and found a honeycomb lying upon the table; and the comb was great and white like snow and full of honey, and that honey was as the dew of heaven, and the odour thereof as the odour of life. Asenath wondered and said to herself, "Is this comb from this man himself?"

And he smiled and said: "Blessed art thou, Asenath, because the ineffable mysteries of God have been revealed to thee; and blessed are all who cleave to the Lord God in penitence because they shall eat of this comb, for that this comb is the spirit of life, and this the bees of the paradise of delight have made from the dew of the roses of life that are in the paradise of God and every flower." (Fife, "Concept of the Sacredness of Bees, Honey and Wax in Christian Popular Tradition," 199–200) such as Eucharist. The story may not have had as much resonance in Italy as elsewhere.

Asenath's shadowy icon has been eclipsed by that of the Virgin Mary, one of the most powerful bee-related icons to have shaped recent civilizations. Religions such as Christianity and Judaism became more influential in Italy, and Christianity did not stress a mythological origin for bees as earlier religions had done. The Virgin Mary, like the honey bee, became a universal symbol of asexual reproduction. Not surprisingly, convents devoted to celibacy and serious prayer to the Virgin Mary sprang up around Italy and Europe. Using the hive as a model of chastity and celibacy, Italian convents for women became centers of education, protection, and community service, even as the Roman Empire disintegrated.

Convents and monasteries in Italy needed wax. Early Christian beekeeping methods involved cylindrical hives like those used by the Hittite, Egyptian, and Greek cultures, and little changed in the type of beekeeping between the ancient societies and medieval Italian society. In examining the Exultet Rolls, which were produced approximately AD 1000, Eva Crane and A. J. Graham emphasized that wax production was as important to convents as honey production: "The Exultet Rolls, which are illuminated manuscripts, clearly show cylindrical hives such as those that were used in ancient Egypt and rectangular hives such as those used in Rome. The length of the hives is usually 2 to 3 ½ times the width or diameter. . . . In general, the proportions conform fairly well to the Roman practice."[19]

The apiary practices shown in the Exultet Rolls are not unlike modern-day beekeeping methods: the parallel arrangement, the flight entrance at one end, the beekeepers standing to one end using smoke. Everything also seems like a typical Roman apiary—except in the center, the Virgin Mary, surrounded by bees.

Early Christian Romans removed impurities from the wax so that candles would burn clearly and art would reflect the colors of encaustic paints (pigmented wax). Until the nineteenth century, when stearin replaced beeswax in 1839, nuns and monks would bleach their wax to make candles.[20] The process of bleaching wax changed according to the location. The Venetians developed a method of bleaching beeswax that, though laborious, produced a superior white wax. After melting the beeswax, they filtered it through a strainer and into a log grinder that pressed the wax into strips. These strips were then placed in the sun to bleach white. The Punics would strain the wax through saltwater twice.

Italy suffered from a series of political setbacks, power struggles, and invasions from AD 400 until the fourteenth century. Yet the desire for bee books and stories persisted, even during the turbulent times. Greek refugees from the Byzantine Empire began to translate ancient Greek texts devoted to beekeeping. The Renaissance brought beekeeping back into focus, at least from a cultural perspective. When Moorish leaders such as Alphonso X ruled, Muslim scholars translated Christian, Greek, and Roman texts, so stories about Deborah, Asenath, and the Virgin Mary managed to survive and spread east, north, and west. While medieval Italy was fragmented and without centralized authority until the rule of Charlemagne in AD 900, the bee-related feminine icons were central to shaping social order within transitional communities.

Christian convents were socially acceptable places for women to be educated and serve the community since there were fewer options for women to marry when the economy turned downward. In the fifteenth and sixteenth centuries, primogeniture laws by which few men inherited land meant fewer men were eligible for women in the upper social strata. According to Judith Brown, "Families with young girls grew desperate to find husbands whose status and honor were commensurate with their own. Their plight is reflected in the well-noted inflation of dowries and in the proportion of women sent to convents as a less costly and socially acceptable alternative to marriage."[21]

Using the convents of San Jacopo di Ripoli and Santa Maria Annunziata delle Murate as examples, Brown argues that women who served as nuns often entered at a very young age, received a good education, and contributed to the society through cultural arts.

The older of the two convents, the Dominican convent San Jacopo di Ripoli, was "a decimated shadow" of its pre-plague self in 1428 when statistics began to be collected. Brown attributes the low number of nuns at that time to the fact that the marriage market opened up significantly after the plague subsided and few women were interested in being nuns.

The Benedictine convent Santa Maria delle Murate selected nuns from the city's most patrician families. As the convent began to increase in size, it admitted *converse* (lay people) to take the domestic pressures from nuns, who could thus spend more time on cultural arts such as gardening, painting, and music.

Theoretically, hives of chaste bees made a perfect model for nuns and convents. Saint Ambrose, the patron saint of bees, who was based in Italy, wrote the definitive text for convents: "Let, then, your work be as it were a honeycomb, for virginity is fit to be compared to bees, so laborious is it, so modest, so continent. The bee feeds on dew, it knows no marriage couch, it makes honey. The virgin's dew is the divine word, for the words of God descend like the dew. The virgin's modesty is unstained nature. The virgin's produce is the fruit of the lips, without bitterness, abounding in sweetness. They work in common, and their fruit is in common."[22]

And convents did good work in their communities, especially during times of plague and other stresses. However, as more aristocratic families faced economic challenges in finding suitors for daughters, convents reflected the increased sophistication of their nuns and the dowries brought with them. Some convents added special rooms for nuns from aristocratic families instead of making them live in dormitory-style houses. In times of medical crisis, Brown conjectures, the reduced access to other people may have kept these convents safer and more hygienic than more populated towns. Brown makes no mention of the health benefits of access to honey, although she does address a common habit of dipping money in vinegar. Since honey has fourteen antioxidants, surely a steady supply of honey may have also made a difference in a time when food shortages and long winters were common. In general, Brown speculates, nuns could expect to live longer than other women because they entered convents at early ages, did not suffer childbirth-related complications, and had reduced exposure to contagious illnesses. In addition, they had daily social networks that included education, art, faith, and other emotional supports that may have been inaccessible to women outside the convent.[23]

These havens for women came under scrutiny beginning in the 1730s in response to Enlightenment principles regarding education and governmental policies that aimed to exert more control over ecclesiastical jurisdictions. The Medici in 1737 began to make convent life less attractive for women, and by 1808, Napoleon had turned "once-bustling convents into prisons and barracks," according to Brown. When convents closed during the nineteenth century, European women had to explore other economic options at a point when societies had few choices. Judith Brown says dismally, "Many exchanged the quasi-familial community of the convent and an honored place in the economy of salvation for risky roles as mothers and dubious places as 'old maids' at the margins of families that had little desire to harbor them in their midst."[24] Yet two nineteenth-century Italian women beekeepers who avoided those fates—Madame de Padua and Madame Josephine Chinni—are often overlooked in bee histories.

Madame de Padua inadvertently contributed to North America's and northern Europe's conversion to the Italian honey bee *Apis mellifera ligustica* (Italian). During Napoleon III's military actions in 1843, the Italian honey bee was brought into northern Europe by Captain Baldenstein, according to Charles Dadant.[25] In 1853 Madame de Padua wrote to Reverend Johannes Dzierzon, who lived in Poland, for a model of his hive.[26] In addition to his other accomplished findings, Dzierzon remains best known for realizing how parthenogenesis worked in the beehive. In many ways, his hive resembled Lorenzo Langstroth's movable-frame hive in North America. Whereas in Langstroth's hive the frames could be removed at the will of the keeper, however, Dzierzon's hive remained more difficult to work because the frames were not hanging but stabilized.

Dzierzon sent Madame de Padua his hive, and as a form of thanks, she sent him a colony of Italian bees, one of the first—and gentlest—colonies to be introduced to northern Europe.[27] Her bees were sent from Mira near Venice via Trieste and Vienna on February 12, 1853.[28] Dzierzon enjoyed extensive connections with the global beekeeping community, and his joy with the Italian honey bee had dramatic ramifications on beekeeping, especially in North America. When beekeepers such as Charles Dadant and Lorenzo Langstroth wanted Italian bees, Dzierzon was the contact person.

Madame Josephine Chinni, a prominent nineteenth-century woman beekeeper located in Praduro-e-Sasso, was visited by the English beekeeper and travel writer Thomas Blow. From his account, she was a very capable queen rearer:

> Here we have a queen-raiser in the shape of the village schoolmistress assisted by her daughter. Madame Chinni, with whom I had dealt for some years, was delighted to see one of her English customers. . . . Madame Chinni likes to manage the affair herself, rather than to employ labour to do it; the result of personal management is, she thinks, far better. A good system of rearing prevails, carried on to a greater extent, by the subdivision of large hives, and further by the use of large nucleus boxes. There were three apiaries—one at the house close to the school, and two on the hills—and the total turn-out of queens would be, probably, not more than between three and four hundred sent principally to England, as Madame Chinni prefers to turn out a small quantity of a good article rather than raise them wholesale. The home apiary was rather small, Madame Chinni having had the misfortune to get nearly the whole apiary of eighty stocks burnt last year.[29]

This rustic simplicity has not changed much from when Blow visited in the nineteenth century, at least to hear contemporary beekeeper Marina Marchese describe her experiences. Marchese is infectiously vivacious and Italian by heritage. She started Red Bee Apiary, which produces organic honey and beeswax products in the United States. In 2009 she traveled to Italy and Sardinia specifically to study nectar sources and terroir of honey. "Who knew there was a city of honey?!!" she writes. "Montalcino harvests 50 types of honey!"[30]

In order to get a better understanding of the nectar sources and honeys of the region, Marchese studied the Tuscan terroir (the climate, soil, and farming techniques of a specific geographic region that impart a unique quality to the agricultural products there). But the medieval commune of Montalcino is known as "La citta del Miele" (city of honey). This, for Marchese, is the Honey Renaissance.

> The samples were endless. A tasting began with a light amber sunflower and we worked our way to the darkest chestnut. They were positively luminous! Sunflower honey was nutty and dark. It reminded me of the miles of fields of golden yellow sunflowers lining the vineyards of the Tuscan countryside. Rosemary honey has notes of camphor and fir. [Our host] explained [that] the dry sandy soil and the dry heat of the Mediterranean climate was perfect for growing. Forest honey was so dark brown reminiscent of burnt toast. It needed a good piece of bread and salty cheese to balance it off. We continued with fica d'india or cactus figs which is the honey from the fruit of the prickly pear. It had a tropic fruit flavor that reminded me of cantaloupe with berries. Not to be missed was the lavender honey. In its full glory delicate and herbal, lavender honey screams for vanilla ice cream or wafers. Spying around the room, I spotted something called Sciroppo di miele al propoli. I had to have it. Syrup with eucalyptus honey and propolis was their own handmade medicine made especially for sore throats and colds. You cannot find these elixirs in the States.

"So much honey, so little time," Marchese wistfully wrote. "I wanted to stay and work there, even without pay."[31]

From watching bees work the lavender, Marchese noted that these were the real Italian honey bees. These bees have inspired its deities and its convents and have become Italy's greatest ambassador, although an overlooked one. In recent years, unfortunately, the Italian honey bees have been under a variety of threats in the United

States and Europe, according to a report published in 2010. Italy has suffered hive losses as high as 40 percent in northern Italy, 30 percent in the southern part.[32]

Even the convents, once centers of bee knowledge, have had difficulties maintaining populations of honey bees in the twenty-first century. Travel writer Grace Pundyk stumbled upon a Carmelite monastery being refurbished and inhabited by five Australian nuns. The Church of Santa Maria del Carmine al Morrocco was constructed in 1459 after the wealthy Florentine Niccolo di Giovanni Sernigi became convinced that a miracle had happened on his land: pilgrims had placed a portrait of Our Lady of Mount Carmel on a tree, on which it would remain embedded, her eyes reproaching him for his lifestyle.[33]

The convent weathered the ups and downs of politics, plagues, and patriarchal posturing until finally it became a crumbling ruin. When Australian nuns arrived in the late 1990s, they set about refurbishing the walls, only to discover that they were working on a Carmelite monastery that houses a nice selection of religious art from the fifteenth to the eighteenth centuries. Unfortunately, it no longer provides shelter for bees. As one nun explained it, "We used to have bees. But they got sick and died. . . . But when we did have those hives, mothers would come to us and ask for some of our honey. They said it helped make their newborns strong. These women came from all over the area. So I guess the honey here was quite famous."[34]

Northern Europe

CONVENTS, GUILDS, AND GOOD WIVES

Ireland was the first non-Roman country to adopt Christian monasticism. Historian David Knowles compared monasticism to a seed that spread slowly, without a designated apostle, starting in the last century of the Roman Empire's unity: "In Gaul in general, where the Roman civilization, half Christianized, was coming to the end in an Indian summer of leisure and literature, the monks were regarded with dislike by bishops and landowners. Nevertheless, the future was theirs."[35] By the fifth and sixth centuries, Irish monasticism had blended the Christian Church's teachings with local Irish clan sponsors.

In these early outposts, Irish women served as abbesses and nuns, with some monasteries serving married couples or both genders. Convents addressed high illiteracy rates by teaching Latin so that people could read biblical texts and translate the messages. The convent or monastery served as the spiritual center of the clanspeople, with land given by a grant from a noble family. Saint Brigit and Saint Patrick were important leaders for establishing the new centers of spirituality, which also became centers of education, health care, and community functions.

Equally important to Ireland's monastic culture was its judicial framework to monitor bee disputes. Ireland had organized a series of laws called Bech Bretha, or "bee judgments." There were nearly a hundred laws, and most were written in AD 438–41. Author Frederick Prete indicates that the Irish laws suggest awareness that the hives had a queen and female bees. The language used in Bech Bretha refers to workers as "victor dames" and a "beo-mothor."[36] Since Ireland had established agriculture long before Christian conversion, apiculture was incorporated into the religious fabric of Irish monasticism, most especially with Saint Gobnait.

During the sixth century, with violence from the clans a constant threat, Gobnait fled County Clare. The Aran Islands, a remote religious colony, offered her shelter and rest. While she was there, according to legend, an angel instructed her to search for a place with nine white deer. Gobnait started on her quest, traveling the south of Ireland until she found three white deer. These deer led her to Ballymakeera, where there were six more deer. She continued on, until she found nine white deer in Ballyvourney. There Gobnait stayed. In 618 she was made abbess of Ballyvourney by its founder, Saint Abban. She was legendary for her accomplishments with honey bees, offering honey and using bees as protection from violence and illness.[37]

Unlike Roman or Asian monasteries, early Irish Christian monasteries emphasized active missionary work, shaped local politics, and acknowledged local fertility festivals rather than banning them. When Gobnait arrived in Ballyvourney, the fertility figures referred to by Irish antiquarian writers as "Sheela-na-gig" (Sheela with the breasts) in the ancient abbey were not erased. "These figures are always female, with legs spread apart emphasizing the vulva, and their 'unusually' crude realism has been hindrance rather than a help in their elucidation," wrote anthropologist Edith Guest in the 1930s. Nonetheless, "survivals of ancient faiths are less remote in Ireland and perhaps nowhere can the significance of these figures past and present be better studied than at Ballyvourney."[38]

For instance, local landmarks outside the monastery suggest an immersion of pagan-Christian beliefs with Saint Gobnait. Saint Gobnait's well is a central stop for religious pilgrims on two days in particular. On February 11, Saint Gobnait's Day, many people visit the well to drink the water. When Guest visited in the 1930s, teacups were placed for the faithful. Whit Sunday, which is observed approximately seven weeks after Easter, also is associated with Saint Gobnait. Explaining the reason that Saint Gobnait is celebrated on two holidays, Guest speculates that the saint blended two traditions: "On 11th February is celebrated the Christian Saint, but at Whitsuntide, the season proper for wonders in the Celtic or pre-Celtic world, the pagan divinity." Whit Sunday is closely tied to baptism rituals of Christianity, so Saint Gobnait's well continues to be a central location for the Irish faithful. Near the well, as Guest describes it, a circular enclosure approximately eighteen feet in diameter suggests an older religion. The entrance is flanked by two upright, flat stones, and "through the narrow entry the votaries must pass, absorbing the mana of the stones by rubbing as they do so." Such older structures convinced Guest that "we are still in sacred precincts though outside those of the Church."[39]

The most common image associated with Gobnait, however, is not the well, as popular as it remains for pilgrims. It is her beehive.

The Bee is more honored than other animals.

Not because she labors but because she labors for others.

st. john chrysostom

Saint Gobnait, Ireland. Courtesy of Patricia Banker.

Gobnait's Clochan, a circular hut, resembled a bee skep and was located a few meters from the church.

A local legend tells of a chieftain who wanted Gobnait's assistance against cattle raiders. As Guest tells it, "Making it a condition that he convert to Christianity, [Gobnait] made a cross above the bee-hive, whereupon the bees became kernes and the hive a brass helmet. The helmet was preserved by the O'Herlihy family, who afterwards became the hereditary wardens of Ballyvourney[,] and [it] was full of virtue for warding off fevers and epidemics."[40] In another legend, Gobnait shook a hive of bees at local raiders and the hive became a bell, said to be identical to a nineteenth-century bell that still existed when Guest was there.

Like local witches, Saint Gobnait was reputed to cast spells on evildoers: "When a thief stole the tools and a white horse of the mason who was building her church, she caused him to gallop round and round unknowingly all night till he was caught in the morning."[41]

The Gaelic word "Gobnait" has been linked to the Hebrew word "Deborah," meaning "bee" and perhaps loosely translated "wisdom with bees," since both women were proactive leaders during times of tribal and clan violence. Other variations of the spelling are Gobnata, Gobonet, Gobnet, and Gobinet. "In no case perhaps is the pagan origin of the cult of the Saint so clearly exhibited as at Ballyvourney," sums up Guest. "We trace the same series of ideas and influences, we may justly deduce that these are what the figure stood for, and today stands for, in the mind of the people. We shall look for the magic water, with its stone, bush, and offerings; for association with cows and other fertility symbols or with witches, Rounds and Patterns. One of other links will generally be absent, but the chain will still be recognisable, often with its sheltering Saint at the end."[42]

This "chain" between Celtic and Christian identities was precisely the lineage that the twentieth-century Irish Arts and Crafts Movement sought to reflect as the movement to seek independence from Britain began to make headway. According to Nicola Bowe, the literary arts from the "Celtic Twilight" (approximately 1890s–1916), needed no introduction. But the visual arts received much less visibility, even though their caliber was just as impressive as that of the literary arts.[43] The Arts and Crafts Movement wanted artists to reflect the rich Celtic heritage, and no one was better at defining new Irish art standards than glassmaker Harry Clarke. When the nondenomi-national Queen's College—now University College, Cork—received a large sum from Isabella Honan's estate, it wanted a chapel that emphasized the Irish heritage. For the Honan Chapel, Clarke designed *St. Gobnait and the Bees,* an intensely stunning window.

Clarke had grown up in his father's art studio, went to London to study, and then, perhaps more important to this book, went to the Aran Islands, where Gobnait had retreated before becoming an abbess. The result, as far as Bowe is concerned, was windows of "an hitherto unseen psychological intensity; a rare feeling for the intrinsic qualities of his materials, and an ability to include symbolic detail without lessening the overall dramatic effect." When creating *St. Gobnait and the Bees,* Clarke distinguished himself with finite Byzantine detail and intense colors. To quote Bowe, his *Wonder-Working Saints* series reflected "an inventive, entirely original, yet contemporary expression of their fundamentally spiritual visions, backed up by extensive reading and research."[44]

Clarke's artistry was acknowledged by his peers. In reviewing the Honan Chapel windows, critic P. O. Reeves gushed: "There is one artist represented in the exhibition who has gone further in achievement than any of his fellows. . . . It is in his stained glass . . . that the full scope of [Clarke's] genius is to be seen. . . . There has never been before such mastery of technique, nor such application of it to the ends of exceeding beauty, significance and wondrousness." Unfortunately, as Bowe suggests, these arts are not impressed on contemporary audiences. Even Clarke himself was pessimistic about the future of stained-glass arts in Ireland in the 1930s.[45] In 2005, while attending Apimondia in Dublin, Ireland, I was struck by the many images of male monk beekeepers but saw no image of or even reference to Saint Gobnait. I heard about her from Robert McLoughlin, an Irish expatriate living in San Diego.[46]

Saint Gobnait's fame and devotion to the Catholic Church extended throughout Judeo-Christian societies, but other convents had their miracle stories with bees. A nunnery located in Beyenburg,

Prussia, achieved legendary status by defending itself with bees during the Thirty Years' War (1618–1648). According to one version, neighboring Swedish knights had laid siege to the nunnery. As the knights attacked, hives of bees were placed around the nunnery so that when the knights tried to take the place, they disturbed the hives, which fell on them and put them to flight.[47] The Swedish forces had a reputation for being especially brutal. Although they did not enter the war until 1640, they laid waste to two thousand castles, eighteen thousand villages, and fifteen hundred towns in Germany alone.

There are plenty of records that indicate honey was traded in northern Europe, but wax was traded only after Christian communities arrived. It is tempting to think that nuns were beekeepers, because we have records of convents paying taxes with wax. At the beginning of Edward I's reign in England, the nuns of Polesworth—now Staffordshire—agreed to bring to the church at Manchester three pounds of wax.[48] In France in 1632, John de Fretar, sexton of the monastery in Chaise Dieu, ordered the surrounding convents and monasteries to pay 600 pounds of marketable wax as rent.[49] Some thirty years later, the monasteries were charging beekeepers 120 pounds of good wax as rent. In *The Antiquities of King's Lynn*, published in 1844, William Taylor says that in the 1400s offerings in wax were common at the priory: sometimes as much as the weight of the person making the offering, sometimes tapers as high as a person.[50]

To meet taxes in a difficult year, convents could hire out a bee earl. Historian Malcolm Fraser cautions that the Middle English Saxon term "beo-ceorl," or "bee earl" or "bee keeper," implies a male keeper.[51] In no way wanting people to mistake a title like beo-ceorl or bee earl for nobility, historian Daphne More clarifies that under the feudal system, the beo-ceorl was a lowly freeman, ranking with the swineherd.[52]

A central concern to Christian beekeepers was swarming, just as it was to Greek and Roman beekeepers. Austin Fife writes of the Christian charms that people used to avoid swarming if possible: "The bees are addressed with the endearing word, 'sizewif,' which is usually translated as 'victory women,' and are asked not to fly to the wood but to sink to the earth and to work for man."[53]

Fife also opines that swarm charms reflected small steps to secularization: "Whereas in the early charms the bees are asked to work for God by gathering wax for the cult, in the later ones, they are asked to bring wax for the Church and honey for the beekeeper's family. Then the Church is often forgotten and the charms only contain a request for the bees to work well for their masters. As a last stage in this process the charms are reduced to the point where one merely addresses them with endearing words and asks them to alight on a limb or on the ground near the apiary."[54]

Other northern European peoples had their own religions in which bees and goddesses were linked to emphasize chastity, moral virtue, and marriage customs. The Polish people, Livlanders and Silesians, worshipped Austēja and her husband Bubilas, sometimes spelled Babilos. According to Patricia Monaghan, Austēja "was the special goddess of brides, who drank honey-mead at their wedding and all other celebrations of their womanhood. Her holy days were in August—busy months for bees, which were gathering harvest honey for their overwintering needs. Austēja was honored with the special form of a human relationship called *biciulyste,* which could be translated 'beeship' (like 'kinship'). Such a relationship was formed between families when swarming bees, following a young queen, moved from one farm to another to establish a new hive."[55]

Some women simply were not cut out for convent life and chose to become "good wives." Once the Protestant Reformation started and beekeeping became less confined behind convent and monastery walls, European women could keep bees in the country or sell wax in the city. European cultures had a code for the good wife, and this code reinforced patriarchal laws. European laws emphasized that rich or poor, women should be kept under the protection of a patriarch, either a father or a husband, if the woman were not ensconced in a convent. Marriage laws were put in place to strengthen that dependency: women could not attend universities, lost their lands upon marriage (an act called coverture), and could not divorce their husbands, even for good reasons such as abuse or infidelity. William Blackstone's *Commentaries on the Laws of England* provides a good definition of coverture: "By marriage, the husband and wife are one person in law; that is, the very being or legal existence of the woman is suspended during the marriage, or at least is incorporated and consolidated into that of the husband; under whose wing, protection, and cover, she performs everything."[56]

European good wives could be beekeepers, but they were not the most educated in society. Books were written specifically targeting this demographic. As Fraser notes, Walter of Henley's *Husbandry*, written about 1250, addresses a good wife "who kept an eye on the bees as she did on the fowls that lived close to her door. She was told how much honey to feed the bees in the winter, and how many swarms, and how much honey should be obtained from each."[57]

England provided a subtler infrastructure for women beekeepers, due in part to the ascension to the throne of Queen Elizabeth I. When Elizabeth came to power in 1558, she influenced beekeepers in direct and indirect ways. Even though churches no longer needed much wax since Catholicism was no longer the official religion, candles continued to be in demand because other forms of fuel were limited or undesirable. Coal was unrefined, and trees in short supply. Because beeswax burned clearly with no smoke compared to tallow or coal, Queen Elizabeth I imposed penalties for the production of adulterated wax. The Wax Chandlers' Company of London had been incorporated by Richard III in 1484. But London had grown by leaps and bounds. More wax chandlers were selling beeswax that was either blended with another substance or had impurities added to make it heavier. Elizabeth declared "an Acte for the true melting, making, and working of Waxe." In addition to being made from "good holsome pur and convenient stuffe," wax had to be stamped with the maker's initials. If the commodity was not pure, the maker was fined, with half going to the queen and the other half going to the cheated party. Since England exported a good deal of wax to its

European neighbors, these penalties raised the English stature in terms of barter exchange. For economic reasons, however, Catholic churches began to accept "false candles" that were covered in beeswax beginning in 1668.[58]

Queen Elizabeth I also set a social standard by drinking honey-based metheglin, or mead. While other European countries had begun to exploit the Caribbean, Indian, North American, and South American lands for sugar cane, coffee, tea, and refined alcohol such as rum, England enjoyed the finest meaderies.

By the end of the seventeenth century, beekeeping knowledge "was becoming more centralized than at any previous time before because it was more readily available than [knowledge about] other insects," explains Fraser.[59] The cultural consequence of knowledge of such basic bee activities as swarming was that thousands of people believed the rhetoric of "hiving off," something that clergymen preached to parishioners countless Sundays, urging them to go to the new English colonies in North America to search for a better life. The hive-based biological realities of a queen bee complicated the political, religious, and cultural standards defining gender during this time. Economics dictated the need for many women to keep bees and participate in the household economy. Compounding the problem was the enclosure of the commons. Disproportionately high numbers of women were forced off feudal lands and into cities. According to Richard Lachman, 40 percent of peasants were landless in 1640 compared to 2 percent in the fourteenth century. In addition, the Little Ice Age, a period of colder and wetter temperatures from 1510 to 1850, caused many cold winters, late frosts, and hailstorms.[60]

As a result, young, single women postponed marriage or chose convents, hoping economic times would change. North American scholar Laurel Thatcher Ulrich comments that from 1650 to 1750 "almost all females who reached the age of maturity [in the New England colonies] married. This trend sharply contrasted to Europe, where during the same period an estimated ten percent of the population remained single."[61]

"Hiving off" to the English colonies was fraught with danger. If English women did not go voluntarily, sometimes they were sent by force. Forced emigration was seen as punishment. Mortality rates were extraordinarily high during the seventeenth century in North America. Economic scholar Robert Jutte found that "in 1618, some forty 'poor maidens' fled one Somerset village alone because they were afraid of being sent to the colonies."[62]

Married women who stayed in Europe were forced to accommodate to the changing social conditions by finding new work. English poor laws rarely acknowledged women's careers or occupations, but Jutte found the most common experience for lower-class women was to have a wide variety of poorly paid jobs. If a woman lived in a rural environment, she could do tasks such as baking, beekeeping, or tending livestock. If she were urban and a good servant, she had a few more resources because she could count on food, shelter, secondhand clothing, and beeswax stubs, which could be sold to wax chandlers. She also might be taught to read, if the employers were progressive.

So seventeenth-century English authors wrote bee books educating women about hives. At least three writers—John Levett, Charles Butler, and William Lawson—wrote books explaining how to keep bees. Of these, Levett's book, *The Ordering of Bees*, was published first, but it didn't receive a wide distribution until 1634. Butler's *Feminine Monarchy* was published in 1609. Nonetheless, according to Malcolm Fraser, Levett's was the first English book that tried to impart knowledge about how to keep bees in the East Anglian, not Roman, countryside. Levett pointedly addresses women: "The greatest use of this book will be for the unlearned and Country people, especially good women, who commonly in this Country take most care and regard of this kind of commodity, although much the worse for the poor bees, because sometimes they want [i.e., lack] help, sometimes diligence, but most times knowledge how to use them well."[63] Levett's book was dialogue-based, following a literary tradition set by the Greeks in which teaching a difficult task is made easier by using a conversational form.

Lawson, in *A Country Housewife's Garden*, published in 1618, declared: "I will not account her any of my good House-wives that wanteth either bees, or skilfulness about them." Lawson offered specific instructions and directions: "The chiefest help she can make of her bees [is] a warm, dry house." Lawson recommends that having forty hives "shall yield you more commodity clearly than forty acres of good ground."[64]

Butler's book *The Feminine Monarchy*, first published in 1609 and then reprinted in 1634, was more well known. Butler built on the structure of Levett's book by making a clearer argument about the queen's role in the hive. He explained that a female rather than a male bee ruled the hive; and in arguing that the hive established a moral "patterne unto men," he set in place a paradigm of women and power

Some Chemicals within a Hive

In previous centuries, beekeepers thought bees responded to human efforts to "call" the bees or "tang" them with pipes and banging pans. Instead, the hive is controlled by a collection of chemicals that govern behavior and respond to environmental stresses: hormones, pheromones, allomones, and kairomones. Hormones are internal chemicals. Pheromones are chemicals shared among the species. Allomones are chemicals communicated between different species, with the receiver being harmed. Kairomones are chemicals communicated between different species, where the receiver benefits. A healthy hive needs all of these chemicals to respond to internal and external environmental conditions.

Since activities within a hive happen in darkness, pheromones are particularly important for communication among bees. There are two categories: primer pheromones that act slowly to set up an environment, such as the presence of a queen calming a colony; and releaser pheromones that prompt bees to react quickly, such as a sting pheromone that causes other workers to sting.

The queen mandibular pheromone is a primer that attracts workers or swarms to the queen and creates a presence of calm if it is distributed. Worker bees "know" that the queen is gone if this pheromone is interrupted even for a brief time, and they will begin the process of rearing another queen.

that still challenges society. Without knowledge of the reproductive patterns of queens, Butler extended many of the ancient Greek and Hittite arguments linking chastity to the hive, especially the queen.

Butler was not always correct in his observations. He thought that worker bees, not the queen, laid eggs. So he also assumed that drones and worker bees conceived. He refers to the queen by a number of masculine titles such as master, governor, king, regent, and prince. Nectar seemed to come from the heavens above, not from flowers, an idea that started with Aristotle.

None of these European authors knew *how* a queen controlled the hive. To them, just being able to observe a queen was a miracle. In Amsterdam, Jan Swammerdam engraved images of a queen under a microscope between 1669 and 1673, although they were not published widely until the 1730s.

To better observe a queen bee's habits, Giacomo Maraldi, an Italian astronomer who moved to Paris, constructed an observation hive with glass in 1712. He assumed that the queen mated in the hive, and he was thus disappointed in his observations. Although much remained to be learned about the queen bee, Maraldi brought many technologies together—coal-fired factories, glass, and microscopes—in his study of queen bees. Maraldi's research led François Huber, working with his wife, Maria, and a male assistant, François Burnens, in eighteenth-century Switzerland, to hypothesize that the queen mated on the wing.

Hannah Wolley could be considered the seventeenth-century Martha Stewart. She wrote *The Queen-Like Closet,* a household manual in which mainstays were honey and wax. Although it doesn't seem as if she was a beekeeper, the book marks the first time that a woman had written about bees for other women. Recipes for metheglin and preserves were included.

In 1672 there appeared a book titled *The Office of the Good House-wife: with Necessary Directions for the Ordering of Her Family and Diary and the Keeping of all such Cattle as to her particular Charge the ever-fight belongs. Also the manner of Keeping and Governing of Silkwormes and*

Honey bees, both very delightsome and profitable. We do not know who the author was, only the initials, F.B.[65]

The author promises to "lay out unto you the ways in how a good House-wife . . . shall order and govern a farm as it may keep and maintain with the Profit and Increase thereof, her Husband, and all his Family." Although the book professes to be written for women, F.B. advises husbands to "make clean the Bee-Hives, and kill their Kings." Later in the text, F.B. sidesteps the issue of gender, stating quite frankly, "I will say nothing of the ingendering of Bees, whether it be by the coupling of Males and Females together, as we see in other kind of Creatures, or by the corruption and rotting of the Belly and Entrails of a young Bullock, whereof Virgil speaketh, but I will describe them as they are already ingendered as what be the properties of such as are fit and like to make good Honey."[66]

The instructions for the reader, assumed to be a "good Woman," were specific and detailed: "The condition and state of the House-wife or Dairywoman is of no less care and diligence than the Office of her Husband; understood always that the Woman is acquitted of Field-matters, in as much as she is tied to matters within the house as the Husband is tied to what concerneth him, even all the business in the Field; and according to our Custom in England, Country women look unto the things necessary and requisite about Kin, Calves, Hogs, Pigs, Pigeons, Geese, Ducks, Peacocks, Hens, and Pheasants . . . and moreover of watching and attending the Bees."[67]

F.B. assumes that the husband will take the honey, if the skep is halfway full. The book advises the good wife "not to kill them [the bees], nor drive them far away if it be possible but keep them to draw more profit out of them afterwards, and when there is no hope of good of them by reason of their oldness; even then you must not use any ungrateful cruelty towards them, as murderously to massacre them instead of recompence."[68]

The vagaries of the Little Ice Age meant more winds and more storms, so shelters called bee "boles" were constructed in rock walls to protect skeps. Contemporary English hive historian Penny Walker points out that most bee boles were located near houses, suggesting that farmers' wives tended the bees.[69] Skeps had a number of advantages over forest-based beekeeping. As a type of hive, they weighed less than other types such as clay cylinders. Rye grasses to weave skeps were readily available and inexpensive. Skeps could be carried conveniently and would prove to be easier to transfer to the North American continent, too.

Whereas Queen Elizabeth I created a legal structure for beekeepers, Queen Anne promoted an economic infrastructure for domestic beekeeping. In 1712 Joseph Warder published the influential *True Amazons; or, The Monarchy of Bees,* which established that honey bees need a queen. "My poor Bees fell again to spreading themselves in search of her," he stated. Warder directly transferred this analogy to his society. The book is dedicated to Queen Anne, and Warder's preface states a dual awareness of power: "Indeed no Monarch in the World is so absolute as the Queen of the Bees. . . . But oh, what Harmony, what lovely Order is there in the Government of Bees!"[70]

Warder makes two cases in his book: too many uneducated people have been keeping bees, and England could have a fine mead-making industry if English people would abolish wine imports from Spain and France:

> These noble Creatures have, of late especially, been much neglected and their Industry not improved in your Majesty's Dominions; the Chief Cause of which has been Ignorance of the right way of managing them, and of the great Profit arising from their Labours, which Defect I have here supplied by Directions at large, gathered from undeniable Experience, which will exceedingly help the Poor, as well as delight the Rich, not only with various Observations and Speculations, . . . but also with a Liquor no ways inferior to the best of Wines, coming either from France or Spain; which if they will but try, they will soon sit down contented under their own Vine, and with me refresh themselves, with drinking Your Majesty's Health in a Glass of such as our Bees can procure us; and no more long for the expensive Wine of our Enemies.[71]

With this as an introduction, Warder aimed specific barbs at good wives and their daughters, who were uneducated and often killed the bees needlessly and unmercifully after the skep was two years old:

"The Age of Bees, and Cause of Death"—says the Good Woman of the House, to whose Protection for the most part the Bees are committed, "this Stock is very good, and heavy, and would stand very well till another Year: But pray consider, it is two Years old already, and if I should let it stand another Year, the Bees will be so old, that they will not be able to labour much next Summer; and now we are sure of a good lump of Honey, that will make us a Firkin of good Mead, fit to be tap'd at Christmas."

Warder offers a reply to the mother in the next sentence:

The Daughters approving of their Mother's Politicks, thus ends the Council of War betwixt the old Woman and her two Daughters, against these her industrious and laborious Servants; and no sooner is this harsh and ungrateful Sentence pronounc'd against these Innocents, but they immediately proceed to Execution; one runs to find a Spade, to dig a Hole in the Ground; another in preparing two or three Split Sticks; a third, the fatal Brimstone Matches to put in them: Thus all Things being prepar'd for an Assault of their rich, but defenceless Castle, they are taken by Storm in the Night, their City plunder'd, and the Inhabitants all slain by Fire.[72]

The graphic destruction Warder describes befell many a skep before movable-frame hives were designed in the 1850s. Almost as if to add insult to Warder's injury, English people did not stop drinking imported wines, and the English beekeeping population remained unsubsidized. Warder's utopia—one in which Queen Anne would rule, the local good wives would practice more humane beekeeping, and the English economy would be more secure—did come to pass, however. England was becoming an empire.

Two good wives became wax chandlers.[73] The Wax Chandlers' Guild was one of the more prominent guilds because of its ability to regulate wax purity and the candle trade. Compared to other guilds, such as those for linen and wool, which were vulnerable to industrialization, wax guilds remained a close-knit craft community. Eliza Bick traded at the Golden Beehive, opposite the Mansion House, London, in 1758. According to Daphne More, Bick's wares included "white and yellow Wax Candles of all sizes, Branch Lights, Winding and Searing Candle, Superfine and all inferior sorts of Sealing Wax, Wafters of all Sizes, Flambeaux, yellow and black Links."[74] It is assumed that she inherited the business from her husband, Edmond Bick, who was listed as a master chandler in 1724.

The other female chandler listed in 1748 was named Hannah Jones, widow of Thomas Jones. Two bills show that she supplied forty-eight pounds of white wax lights every month to the Duke of Bedford and that in 1778 she sold forty-two white wax lights to Magdalen College, Cambridge. For thirty years, she had participated in this guild.[75]

Feminine leadership in the beekeeping arts was spread across Europe. Following in the steps of monarchs such as Catherine the Great of Russia and Elizabeth I of England, Maria Theresa, sovereign ruler of Austria from 1717 to 1780, became a highly visible patron of beekeepers. She started the first beekeeping school in Europe in 1766.

A Swiss scientist, Cátherine Elisabeth Vicat, designed her own hive, which differed from a skep because it extended horizontally instead of vertically, as is now the custom when adding supers to hives. This type of addition would have made it easier for her to physically manage the hive.

Vicat was the wife of an aristocratic Swiss academic and is considered one of the first female bee scientists.[76] She wrote extensively about the bees' need for cleansing flights. Her papers were published by the Berne Society. From 1760 the Economic Association published sixty-five reports and notes of her experience. Even in the United States, people were quoting her research to understand cleansing flights, although her work was not written in English.[77]

Another Swiss woman, Maria Aimée Lullin, made bee history by being stubborn. The daughter of a Swiss magistrate, Lullin waited seven years to ask her sweetheart, François Huber, to marry her. Born in 1750, Huber had suffered from weak eyesight since he was a young boy. Thus, he was completely blind in 1765, but he had the good fortune to meet Maria at a dancing class. William Artman and L. V. Hall write, "Fearing that the loss of sight might unfavorably affect the dearest object of his affections, he resorted to dissimulation. While he could discern a ray of light, he acted and spoke as if he could see perfectly well, and often beguiled his own misfortune by such pretenses."[78]

Maria's parents opposed the couple's marriage, even though Huber was born into a wealthy family. But Maria overcame her parents' authority by waiting until she was old enough to propose to François; she had to be twenty-five years old according to Swiss law. Maria's uncle walked her down the aisle.[79]

Huber was remarkably perceptive. François Burnens, a peasant with no formal education, carefully observed Huber's beeyards and, based on Huber's conjectures, applied Huber's theories in the field, such as preventing a queen from leaving the hive. This led to the realization that queens mate in the air. Following in the footsteps of a scientist named A. G. Schirach, who figured out that queen cells could be reared from larvae in worker cells, Huber showed that when a virgin queen was confined, she would lay only drone cells. Not only did he establish that queens determined the sex of their eggs, but he and his collaborators also began to do basic queen rearing: "Huber . . . used the method of transferring larvae from workers cells to queen cells which is now used almost exclusively to rear queens commercially, the only difference being that he transferred larvae to natural queen cells rather than to artificial queen cell cups."[80]

Also on the Huber team, a female naturalist only known as Mlle Jurine dissected the honey bees' wax glands and observed ovaries in all the workers. Jean-Pierre, the Hubers' son, aided the team as well. Of his collaborative process, Huber once said to a friend: "I am much more certain of what I declare to the world than you are, for you publish what your own eyes only have seen, while I take the mean among many witnesses."[81] Maria faithfully transcribed her husband's theories about all of these experiments into prose.[82]

In 1792 *New Observations on Bees,* addressed in the form of letters to naturalist and illustrator Charles Bonnet, was published. The book that resulted from this collaboration set the stage for nineteenth-century developments of the Langstroth movable-frame hive and bee space.

Unfortunately, the Hubers were indirect victims of the French Revolution. The turmoil caused by the political uproar disrupted banking channels on which the Hubers depended, and because of financial distress, Huber had to release Burnens. In a letter to Jean André de Luc, he wrote, "I might have got some interesting and useful results if I had not been forced to discharge Burnens as I was no longer able to pay him or to feed him and thus to lose my eyes a second time."[83]

When Maria died, Huber spent the evening of his life at Lausanne under the care of his daughter, Madame de Molin. He never really recovered the joy he shared with his wife, however. Huber once said of Maria, "As long as she lived, I was not sensible of the misfortune of being blind."[84]

Wax modelers such as Madame Marie Tussaud and Catherine Andras, "Modeller in Wax to Queen Charlotte," indirectly shifted public appreciation for the artistic benefits of beeswax. While its properties have always been appreciated for smoothing dry skin and polishing unvarnished wood, these two women raised beeswax to the medium of high art in the eighteenth century. Since beeswax resembles the sheen of human skin, it has been the perfect medium to preserve likenesses of political leaders and celebrities. Roman artists made busts of famous leaders. Life-sized effigies of kings, queens, pets, and profiles have lasted through the centuries. But Tussaud's models and Andras's sculptures generated public discourse on race, politics, and class differences before visual technology existed.

In an era devoted to technological advances, beeswax served as a good vehicle for creating anatomical structures for study. Thus,

Marie Tussaud, born Anna Maria Grosholz, grew up working with a Swiss doctor named Philippe Curtius, who created wax figures to study anatomy. His molds were so good that people wanted to view the figures, so he opened a small display, his Cabinet de Cire, in Paris in 1770. Having started under Curtius's tutelage, Tussaud made her first mask of Jean-Jacques Rousseau, and she later created masks of Voltaire and Benjamin Franklin. Her art was so admired that the French monarchy invited her to teach wax arts to the family.

When the French Revolution started, Tussaud was arrested and jailed. Her head was even shaved for certain execution. The only thing that saved her was the intervention of Jean-Marie Collot d'Herbois, an artist and political revolutionary who sat on the Committee of Public Safety. Perhaps the close call was enough to make an impact on Tussaud, for throughout the Reign of Terror, the death masks of the victims of the guillotine do not reflect fear. For all the destruction happening around them, the masks of Louis XVI, Marie Antoinette, and Robespierre suggest acquiescence and peace.

Marie married François Tussaud when she was thirty-four, but they lived together for only seven years. She eventually settled in London, and according to More, began the next stage of her career under less desperate circumstances. Opening a museum with the model sculptures she had brought from France, Marie and the Tussaud family continued her art for four generations. The models are still made in similar ways:

> The unclothed parts of the figure, usually head and hands, are made from 75 percent beeswax with 25 percent of Japanese wax, a vegetable product. The figure is sculpted in clay, then a plaster mould is made from the head and tinted molten wax is poured into it. The head is hollow so that a hand can be inserted to fasten the carefully matched glass eyes, and real hairs, about three hundred to a square inch, are put in by hand. The final coloring of the face is done with water colors. Concealed parts of the body are cast in fiberglass nowadays, but even in her later years, Mme. Tussaud visited the death cells to take particulars of condemned men.[85]

In an age before cosmetics and conspicuous consumption, Tussaud offered the public a chance to admire the permanence of art. Her museum gives people a temporary respite from political turmoil and a chance to be entertained while also being educated.

A contemporary of Madame Tussaud, but not nearly as famous, royal wax modeler Catherine Andras was an exceptionally talented artist. Her work still hangs in the British Museum, and her subjects ranged from Queen Charlotte to Captain Horatio Nelson. In sharp contrast to her courtly contemporaries, Catherine Andras did not try to hide Queen Charlotte's mulatto heritage in her beeswax sculpture. The court had already been surrounded by writers and philosophers who promoted abolitionism by including Charlotte's African heritage in their writings, but no one had sought to reflect Charlotte's racial features in sculpture until Andras's work was unveiled. King George III and Queen Charlotte moved the English aristocracy to an abolitionist stance, and chattel slavery was abolished by 1815. Just as Madame Tussaud "humanized" the French Revolution, so too did Catherine Andras humanize Queen Charlotte. In an age before cameras, videos, and more sophisticated technology would become a "lens" for social justice issues, beeswax became a political tool in the hands of these two women during tumultuous times.

Other European women artists would use their arts in the literary world, appropriating the honey bee to suggest moderation. Since Aesop's fables had been translated and were available for eighteenth-century European audiences, order and reason could be learned in the literary forms used by the classical Greek and Roman writers. (The chief criticism of eighteenth-century literature is its heavy-handed didacticism.) Considered part of royal society as a result of her mother's literary endeavors, a French lady named Madame Antoinette Thérèse Deshoulières wrote a fable about a king who is elected to serve the people, rather than the other way around ("La Police des Abeilles," published posthumously in a collection in 1778).[86] This was precisely the course of events that would play out in the United States.

One French woman would have done well to read the fables and practice the lessons of moderation. Anne Louise Bénédicte de Bourbon, duchesse du Maine, was one of the most flamboyant figures of the eighteenth century, and she earned the negative connotations of the title "queen bee." She was born in 1676, the granddaughter of a famous general, and married Louis August de Bourbon, Baron de Sceaux, a member of the French nobility. In 1703 she founded the Order of the Honey Bee.

While her husband was in Paris, Anne Louise Bénédicte de Bourbon devised elaborate neoclassical *tableaux vivants* in which she herself took the role of the Queen Bee, the knights being her faithful workers. According to bee historian Bodog Beck, "The Order was an elaborate parody of chivalry. Pompous reunions became an established and permanent amusement of her court." Beck seems as charmed by this pageant as the knights, calling her at one point in his article "some Ariel-like creature of the skies."[87] The medal that the knights received read: "Piccola si ma fa pur gravi le ferite"—"I am small, it is true, but I can inflict deep wounds." But as word leaked out about these elaborate parties, most of which were paid for by peasants who could barely eat, her husband lost his standing in the French court. The Order of the Honey Bee became associated with extravagant overspending.

Other public shows regarding women and bees took place in England. The Astley Amphitheatre of Equestrian Arts started in 1768. Philip Astley, the "Father of the Modern Circus," assembled talented equestrian riders, acrobats, clowns, and jugglers. The act that exceeded them all was a bee charmer. Philip's wife, Patty, won the crowds' hearts by riding a trick pony among a swarm of bees.[88] As described in Ricky Jay's book *Extraordinary Exhibitions,* Patty Astley circled the ring on horseback with the insects covering her arms like a muff. There is no word about the number of stings she may have inadvertently suffered in the process. The Astley Circus provided the template for future successful entertainment, but no one tried to follow Patty's bee charmer act.

European women beekeepers were steady contributors in small ways to the scientific, agricultural, political, religious, and economic circles that had formed during the Enlightenment. The caste system of queens and workers created social challenges, but women bee-keepers were much more visible in Europe than on any other continent. Women were able to position themselves in various careers, creating value-added products such as beeswax or being beekeepers. The social codes dictating feminine behavior had to become more flexible after centuries of unpredictable changes in weather, political and religious turmoil, expanding population, and industrialization. Honey bees and the European women associated with them served as a type of social stabilizer, symbolizing reason, moderation, and order, even as global politics exhibited anything but these qualities.

NINETEENTH-CENTURY WOMEN BEEKEEPERS

European beekeepers were hampered by the limitations of skeps, swarming and the inaccessibility for ready maintenance. In 1851 the Langstroth movable-frame hive, designed with bee space and movable frames, made beekeeping much more manageable. Yet, as far as women were concerned, the ideals of the good wife and the convent nun were being replaced by the Cult of Domesticity, and the new ideal was the "Angel of the House," a fictional woman in a poem written by Coventry Patmore. With industrialization in full swing in Europe, women were shifting to more organized schedules and, in some ways, more gendered roles. The angel of the house was defined by four traits: domesticity, piety, chastity, and submissiveness. No longer was being a good wife, with its rustic implications and barnyard responsibilities, considered an ideal. With new standards of living brought by industrialization, women were expected to mind the home but not necessarily mind the bees.

Samuel Bagster's *Management of Bees with a Description of the "Ladies' Safety Hive"* was the right book for women at the wrong time. Published in 1834, it appeared in Europe before the concept of bee space was understood and the movable-frame hive was designed. For contemporary audiences, the manuscript is a quaint curio. But Bagster's book is a love letter opened to the world. In a world of patriarchal rules and judgments, this book sheds light on a very real friendship binding a nineteenth-century marriage.

Bagster begins with a bold and forthright invitation: "The author desires to invite ladies to the study, on his assurance that they need not fear either a sting or ingratitude; their attention will be appreciated by the sagacious little fellows." Bagster's wife is never referred to by name, but she is as clear a presence as the bees. He clearly intends for this book to ease her labor: "I felt it was my pleasure to save her as much annoyance as possible while pursuing her daily avocations. But the constant fear of being stung, or not managing the bees correctly so influenced my partner that she confessed her fear and begged to decline the duty."[89]

Not being a lazy sort, the wife continued with her requests for her husband "to find bees that would not sting" and a hive that could be properly managed without angering the bees. Bagster could not find bees that would not sting. But having an inventive mind, he turned his attention to hive design. He willfully admits that his hive may not suit his fellow countrywomen, but he asks that they acknowledge "I have done what I could to smooth some of the hindrances to this study, under the best feeling of a married life—a persevering endeavor to please my wife." In what surely is the best advice ever given a beekeeper, regardless of gender, Bagster writes the following words: "Courage is not half so necessary as judgment: an old proverb has it: He that fights and runs away / May live to fight another day."[90]

Bagster's book contains charming tales of his mistakes with the bees, his successes with a queen, and his efforts to engineer a hive that improves on others by providing a cool storeroom, easy ventilation, protection of bees, and simple management of honeycombs. He even admits to putting a thermometer in his hives, although the bees filled it with propolis. Bagster could not have foreseen that the Langstroth movable-frame hive would make his book irrelevant, but his book remains one of the more entertaining how-to guides for women.

The Swiss scientist Cátherine Vicat's collateral hive design continued to be published extensively in England in Robert Huish's nineteenth-century book *A Treatise on the Nature, Economy, and Practical Management of Bees*. As did Warder's book *The True Amazons*, Huish's advocated domestic beekeeping as a way of ensuring Britain's independence from foreign markets even as the country expanded its empire. Huish promoted Vicat's hive as easier than using skeps.

In direct contradiction of Huish, Scottish author Clementina Stirling Graham worked with her gardener in Duntrune, Dundee, to promote skeps. Her translation into English of French author Jonas de Gelieu's book *The Bee Preserver* in 1829 may have influenced her in this. Whereas other European beekeepers were moving toward fixed-comb hives such as Vicat's, Graham never lost her affection for skeps. However, because Graham's skeps were subjected to harsh weather conditions such as storms and strong winds, she "had the four south-facing bee boles built in her garden walls."[91] Perhaps this extra bit of shelter encouraged her to keep bees for sixty years. Rather than going out with a whimper, at the age of ninety-six Graham became a founding member of the East Scotland Beekeepers' Association in 1876.

An equally commanding woman in English and Irish beekeeping was Baroness Angela Burdett-Coutts, president of the British Bee Keepers' Association (BBKA) from 1876 to 1904. She inherited much of her money from her grandfather Thomas Coutts, a banker, but she never rested on those bank accounts. In 1871 she became the first woman to be raised to peerage as a reward for services to her country. Burdett-Coutts chose not to marry until she was sixty-seven, and then to a much younger and kind man, William Ashmead-Bartlett, an American. He worked with her to promote her social causes. In an interesting reversal of custom, her husband changed his name to Burdett-Coutts and then joined his brother as a member of the Westminster Parliament.[92] By the time she became involved with the BBKA, Burdett-Coutts had helped set up schools for the poor, fund churches in Africa and Asia, bankroll entire sections of the Irish during the famine years, and restructure some of the worst neighborhoods in London while compensating generously those who relocated. In these projects, she sought ways to assist the poor. She also took a stand for those who could not speak—insects, animals, and children—and she set up societies to prevent cruelty to living organisms. Eva Crane quotes T. W. Cowan in describing the baroness: "She elicited a joyous hum from those who had so long been queenless."[93]

Burdett-Coutts believed beekeeping to be a noble hobby, and she worked on behalf of both Ireland and England to promote beekeepers. In fact, the Irish Beekeepers' Association, and the separate Northeast of Ireland Beekeepers' Association, began as a "result of a BBKA mission to Ireland."[94] In England, Burdett-Coutts worked to bring a honey business to the Columbia Market; this ultimately failed, but it was a mark of her charity and ingenuity at developing socially sustainable venues for beekeepers and poor people. She tended to be a one-woman show, however. According to Karl Showler, the national council of the BBKA remained an "all-male preserve to the end of the century."[95]

TWENTIETH-CENTURY WOMEN BEEKEEPERS

As two world wars redefined agricultural extension policies, European women took advantages of the vacancies created by the vast migrations to the United States and Australia. They organized for the right to vote and better working conditions. Women beekeepers attained leadership positions in beekeeping associations, bee magazines, and research laboratories. This leadership consequently set up the expansive research that has been carried out in science, history, and extension in the twentieth and twenty-first centuries.

An anonymous book titled *Instruction in Bee-Keeping for the Use of Irish Bee-Keepers,* published in Dublin in 1912, signaled a cultural shift and acceptance of women working outside the home. Women were encouraged to be beekeepers because "to work [bees] to the greatest advantage, it is essential that the right thing should be done at the right moment; hence, bee-keeping is especially suited to cottagers and small occupiers who are not likely to be absent from their homes for several days at a time: it is also an industry for which women are well-adapted; many of the most successful and capable bee-keepers are women."[96] Similarly, James Watson's *Bee-Keeping in Ireland,* published in 1981, lists a number of women in positions of leadership, including Mrs. Annie Cronin of Kilgarvan, listed as an expert in the

Irish Beekeepers' Association; Miss R. Shackleton, pictured beside her hives; and Baroness Prochaska, elected president of the Irish Beekeepers' Association in 1921.[97]

Beekeeping laws have been on the books since antiquity, but rarely have women been guilty of transgressing them. However, in Ireland on June 30, 1901, Mrs. Margaret Daly was sued by John P. Nolan because she "deliberately open[ed] her beehives and allowed her bees to escape to do Mr. Nolan serious bodily injury, and with having done it purposely to prevent Mr. Nolan from attending his work."[98] The beekeepers rallied to her support, with some paying part of the punitive damages Mrs. Daly acquired.

The advent of World War I would distract beekeepers from these types of domestic transgressions, however. Both World War I and World War II inadvertently created a leadership vacuum among bee associations that opened opportunities to European women beekeepers. While men were occupied in military endeavors, women took advantage of educational and occupational opportunities that had previously been closed to them. Karl Showler documents the change in this way: "The great change came after the First World War because the loss of men by death or wounds removed a potentially active generation and also during the war many women entered regular employment and were ready to play an active part in association affairs."[99] Patriotism could be shown by providing beeswax and honey for troops and medicines. The severe sugar rations meant many Europeans turned to honey for some variation in their diets. Wax was used in a number of ways, from waterproofing boots and tents to waxing down planes to get better fuel mileage.

It therefore makes sense that women had extensive knowledge of bees because of their work during World War I. A. B. Flower published a book titled *Beekeeping Up to Date* in 1925. In her tenure as a member of the British Beekeepers' Association (BBKA), she earned the lecturer's and First Class Expert certificates. Eschewing conventional modes of transportation such as cars and trains, Flower rode a motorcycle to visit beekeepers.[100]

Flower set a paradigm for Annie Betts, editor of *Bee World,* to follow. Born in 1884, Betts began beekeeping in 1900. Betts did not intend to be a beekeeper, however, but rather an aeronautical engineer, "a most unusual career for a woman in those days," to quote Showler.[101] In her retirement address to her readers, she forthrightly stated that she wanted no memento "of the many years I have spent in work which, though interesting, was not of my own choosing."[102] She stepped into the editorship of *Bee World* amid financial difficulties with the sudden death of Captain James Morgan.

In 1919 Betts developed a very close friendship with Ahmed Zaky Abushâdy, an Egyptian doctor of medicine. Abushâdy had come to Britain as a medical student in 1912, studying at Oxford. He began a company with two hundred hives and traded in bee equipment. He was a firm believer in standardization of equipment. In 1919 Abushâdy formed the Apis Club and started *Bee World.* The Kent Beekeepers Association started *Bee Craft* a little earlier the same year. Betts described the doctor as an international pacifist and used the Apis Club as a foundation for the BBKA.[103]

With such a strong alliance in international circles, Betts set high standards for *Bee World.* She navigated the publication through difficult times, most notably two world wars, and continued Flower's tradition of riding to apiaries on a motorcycle. As an editor, Betts was forthright and opinionated. She had no patience for biodynamic plans of beekeeping such as Rudolf Steiner's. She also politely but firmly rejected any notions of a "pure" race of German bees when reviewing the German journal *Archiv Fuer Bienenkunde,* which promoted the possibilities of such. In the long run, she wrote, "*the proof of the strain is in the honey crop* given certain other essentials such as good wintering and temper not utterly demonic. We can safely reject the Nazi ideas of 'race,' which are obviously biologically unsound."[104]

But Betts also praised her colleagues and was not above taking her own self to task when she fell short of her own standards. "Sackcloth and Ashes!" reads one title. She writes of herself, "The Editor feels that she ought to sit in the above. . . . All of us make mistakes, but I know of no other research worker who would acknowledge a mistake so unreservedly, except perhaps François Huber."[105]

Against impossible odds, Betts kept *Bee World* going almost single-handedly through World War II, finding other people had no time. In 1949, when Betts was sixty-five, she chose to retire from her position as editor. At this point, she was so deaf that she had great difficulty communicating with people on the telephone. But the handicap was not the primary reason she wanted to retire. As ever, Betts was blunt about her primary reasons: "I am tired of reviewing others' work and doing little original work of my own, and of acquiring the reputation of a narrow specialist—a species of human being I do not admire."[106] Perhaps this is why, when O. Morgenthaler wrote an obituary about Betts, he invited other people to express their experiences, for he had found that she was reluctant to discuss personal friendships. "Miss Betts misunderstood me—I believe intentionally— in order to keep from writing about herself," he said.[107]

When Betts stepped down, she left *Bee World* on solid financial footing, which made Eva Crane's transition to editor much easier. "Although there was a remarkable expansion of beekeeping in the years of critical food shortages following the Second World War, this did not produce a flow of books from female hands," pondered Showler.[108] Surely the stress of the war years, the years of rationing paper, difficulties of transitioning to a peacetime economy, and the exceptional challenges facing beekeeping were why there were not more books published by women during the years following World War II.

In spite of the turmoil ensuing from wars, economic depressions, and infrastructure creation, women were making scientific advances regarding bee nutrition and flower pollens. Born to a Polish father, Dr. Anna Maurizio continued a tradition of Swiss women bee scientists begun with Cátherine Vicat and her hive. In 1928 Maurizio joined the Bee Section of the Swiss Federal Research Institute for Milk Husbandry. Her doctoral thesis was on mycology, and she started bee research with a study on fungi that were potential honey bee pathogens.

Anna Maurizio in her lab at the Liebefeld Institute. Courtesy of Joseph Hättenschwiler, photographer.

Maurizio contributed other findings, concluding that the natural length of a life of spring–summer and autumn–winter bees is different.[109] In the 1960s, she was working on pollens obtained by different bees, finding that "many of the bumble bee honeys contained red clover pollen, but there was also a great variety of other pollens, showing that bumble bees visit many of the food sources available to them, and not only those designated 'bumble bee flowers.'"[110] In summing up Maurizio's career, Eva Crane stated, "From the 1930s until the 1980s, Dr. Maurizio published pioneering papers on the quantitative pollen analysis of honeys, and her research led many countries to adopt this investigative method for determining plant sources of their honeys."[111] Maurizio's accomplishments were a bright spot in a field that not had gained much direction since budding palynologist Christian Konrad Sprengel had published his findings in the late eighteenth century.

However, of all the prominent women in bee research, Eva Crane outshone everyone in terms of sheer magnitude of research. Her *World History of Beekeeping and Honey Hunting* (1999) is still the

Eva Crane checking a beehive, England. Courtesy of International Bee Research Association.

definitive text on global bee migrations, technology, and culture. Her other major works, *Bees and Beekeeping: Science, Practice and World Resources* (1990) and *The Archeology of Beekeeping* (1983) continue to be equally informative. But her lesser-known volumes *The Rock Art of Honey Hunters* (2001) and *Making a Beeline* (2003) are just as irreplaceable in my bee library, as are a portfolio of editorials from *Bee World*. Her clarity of voice, straightforward prose, and genuine enthusiasm for all beekeeping forms around the world still inspire me. While I was attending Apimondia in Ireland in 2005, I arranged a trip to visit Crane in England.

There was nothing easy about going to see Eva Crane. Train

traffic had been rerouted. Rain fell incessantly. *The Rock Art of Honey Hunters* was in my backpack.

Penny Walker, Crane's friend and associate, kindly offered to pick me up at the train station. She had cautioned me that Crane was accustomed to her afternoon naps and that probably an hour at the most would be all that I could have to talk with her.

I also had been counseled by Richard Jones, who was then director of IBRA, that Crane did not suffer fools gladly. I had already experienced some of her impatience when I called to make an arrangement to see her.

"I am writing a book about women and bees," I explained.

"Why do you want to do that?" she immediately countered. "Is there really such a need?"

"Well, I didn't write as much about women as I had wanted to in *Bees in America: How the Honey Bee Shaped a Nation,* my first book."

"Did you write about all the bees in North America?"

"No," I said, "just the United States."

"Well, why didn't you title it *Bees in the United States*?"

Suddenly, I did not care to interview Crane at all, much less go through the logistics of arranging plane, train, and taxi travel to see her.

She offered, rather suddenly: "I do not know that I can help you, but I will be glad to meet you if you can come to England. I tire easily, so it can't be for a long visit."

So, I found myself knocking on her door at Woodside House, prepared with small gifts for her canine companion. I had had a day of just-in-time dashes through subway doors and safety checks at the airport terminals. Only optimism and adrenaline were carrying me into the conversation. Being an accomplished traveler herself, Crane may have sensed the abrupt wordlessness that happens after a series of "stop and go" lurches toward a goal.

"Would you like some tea?" Crane offered immediately.

"Would love some," I blurted. We talked about her Irish property. She signed my copy of her *Rock Art* book and laughed when

I explained I could not fit *The World History of Beekeeping and Honey Hunting* in my suitcase.

"No, I suppose it would be . . . inconvenient," she smiled.

Very gingerly, we talked about this manuscript focused on women and bees.

Even today, I marvel at the places Eva Crane went, chronicled in *Making a Beeline*. Her husband gave her a great deal of flexibility, and she traveled to many places alone. As she described in *The Rock Art of Honey Hunters,* Crane had planned to visit a shelter near Toghwana Dam, Zimbabwe, but could not approach it because the shelter was occupied by armed insurgents. When I asked if she was ever concerned about the places she had gone, she used the word "apprehensive." "Yes," she said thoughtfully. "I have been apprehensive."

"But," she continued carefully, "You must trust your mind and trust the bees that brought you to that place."

"I do not think there is a need for a book to focus exclusively on women beekeepers," she stated bluntly.

"That may be true," I conceded out of weariness and politeness. "But I want you to know this: that when I was writing my book, the *World History* was the one book that gave me an example of what a bee book could be."

We were quiet again. The rain had stopped, and the sun lit up her garden. The sunshine is what I remember most, it forming a backdrop to the picture window in the library and softening Crane's posture in the narrowback chair. In the brilliant, sudden sunlight, Crane sat surrounded by a comfortable clutter of books, tea cups, and rugs and an air of solid accomplishment.

In making our goodbyes, I was actually looking forward to my return to Dublin and to the solitude that engulfs me when traveling solo. The layer of kindness between Crane and me was enough; there was simply no time or need for more than that.

As I was reaching for the door and turning to thank Crane for her hospitality, she offered a brief, tense hug. "Remember to trust the bees," she said firmly and with clarity, and I have done just that.

When Crane became a beekeeper in 1941, she began a lifelong journey that included the entire world. She was, simply, the most important focal point to shape apiculture around the world, using apiculture as a tool to unify a fragmented post–World War II Europe. Crane also fundamentally changed approaches to international beekeeping by respecting international traditions, and she brought a humanitarian approach to extension efforts in less-developed countries.

But as with many other beekeepers, Crane had chosen another career before apiculture. Born June 12, 1912, Eva Widdowson Crane grew up in Dulwich, South London. She was educated at Sydenham, a girls' school that provided dedicated teachers and quality education after World War I. She won a scholarship to read mathematics at King's College, London. One of only two women then reading mathematics at the university, she completed her degree in two years. In an e-mail exchange with close confident Richard Jones, he stressed: "She saw nothing strange in the fact that she was one of only two women on the course. All that mattered to her was the intellectual challenge and the intellectual rigour applied by those around her."[112] Widdowson earned an MSc in quantum mechanics and received her PhD in nuclear physics in 1938.

From 1936 to 1940, Widdowson was a lecturer at Hull University before accepting an appointment in physics at Sheffield University. To quote Jones, "a burgeoning career in the rapidly developing world of nuclear physics seemed assured."[113] In 1942 she married James Alfred Crane, a stockbroker who was doing wartime service in the Royal Naval Volunteer Reserve. When they married, they received some beehives as a wedding gift.

Unlike Miss Annie Betts, Crane walked toward apiculture, charting new territory as she and her husband became better beekeepers. When Betts stepped down as editor of *Bee World* in 1949, Crane stepped into the editorship. The "business worries" that Betts had faced had been resolved. Deciding that more awareness needed to be brought to international beekeeping, Crane also founded the Bee Research Association in 1949—the word "International" was not added to the title until 1976—and was appointed director. In her first *Bee World* editorial in 1950, she emphasized the importance of the magazine because "the subject is expanding so rapidly and in so many directions."[114] As director of IBRA, Crane often hired mothers. When I asked her about her workforce, Crane responded as if it were obvious: "Mothers are the best multitaskers."[115]

After Crane retired from the directorship in the mid-1980s, she continued to travel and published another 130 articles and books. In addition to the most accurate scientific observations available at the time, Crane offered astute political statements on apartheid, gender discrimination, and the Cold War. The following is just one observation: "A few years ago, when I was in Austria, I happened to visit an apiary a few yards from the Iron Curtain, beyond which lay Hungary. I watched the bees flying to and fro across the barbed wire that formed the boundary and they, of course, did not understand that it *was* a boundary."[116]

Crane died in 2007 while I was surveying Peabody coal-mine sites in Mackey, Australia, but for me, Crane's research is embedded in my own. Her article titled "Letter from Australia" was in my backpack while I was driving in the outback. Her research is woven into every chapter of this book. Even today, I picture her sitting in her library. Just like the bees in Austria, Crane's words fly to and fro from her pages to mine, and while she has crossed a boundary, I am not cognizant of it.

BEES FOR DEVELOPMENT

In many ways, Nicola Bradbear has become the most visible and prominent leader in rural development bee programs since Eva Crane stepped down from IBRA. With her leadership, Bradbear has shaped international beekeeping extension efforts in ways that soften some of the oppressive ideas that dominated prior efforts. Her organization, Bees for Development, is a relatively recent association. Bradbear explains: "I started work at International Bee Research Association in 1983 when Eva Crane was in charge. She 'retired' as Director in 1984. . . . By 1993, an executive decision had been made that IBRA would not 'do development' any more."[117]

In very plainspoken fashion, Bradbear expounds: "Helen Jackson, who started as my assistant when IBRA was moved from Gerrards Cross to Cardiff, and I were made redundant." Bradbear and Jackson started Bees for Development the next day, April 1, 1993, and Bradbear fortunately obtained an apiculturist lecturer post at Cardiff University. That position lasted until 1996, after which she devoted herself full time to Bees for Development.

"There are two separate jobs," explains Bradbear: "providing the services of Bees for Development, and raising funds for the work. These are two quite separate tasks and it is very difficult to raise funds, although there is huge demand from developing countries for our information!" The best-known international extension service of its kind, Bees for Development continues to be the go-to place for information regarding beekeeping in undeveloped areas.

Interesting too is Bradbear's support for bringing Chinese honey into the market. Bradbear will not condemn Chinese honey, in large part because she sees enormous opportunities if the communication between China and other industrialized countries can stay open.

Bradbear's position on honey contamination is not shared by everyone, but it is certainly valid. The European Union's honey standards are stringent by any country's standards, but they are transparent and, at times, attainable with assistance. In setting these standards for Africa, which contains many third-world countries, there may have to be some flexibility in addressing darker honeys, since much of the world has a preference for lighter honeys. As far as Africa is concerned, Bradbear practices a concept called "tough love" in Grace Pundyk's travel narrative *The Honey Spinner.*

"What the African countries need is information so they can work out how to get around the issues," explains Bradbear. "We've held workshops to encourage marketing and we also helped start ApiTrade Africa. We're hopeful that by forming such a bloc, member countries will be able to work towards launching African bee products into the global market."

"Beekeeping reflects the huge disparities between rich and poor nations. The whole scene is based around the European honey bee and beekeeping methods. But there's a whole world out there that practises otherwise, and still manages to produce delicious honey," Bradbear summarizes.[118]

In the case of China, which is striving for world-leader status and has an abundance of light-colored honeys, working with beekeepers to meet these international honey standards is to everyone's benefit. In her role as director of Bees for Development, Bradbear brings a different perspective to this volatile discussion: "All you need is one [bad] batch of Chinese honey and then all Chinese honey is banned. But it's not all bad honey. It's certainly not all contaminated. And of course, it's cheap. Labour is cheap in China."

Bradbear brings much practical experience to the issue of international honey production and can see many opportunities to improve the process of EU legislation and perhaps the lives of those who keep bees outside of its influence.

I am one such person. In Ireland in 2005, I attended many of the Bees for Development panels. Many topics concerning rural beekeepers were more approachable than the science presentations. The speakers interacted with the audiences, and fewer presentations were driven by technology. However, there was not one presentation about the United States. I was struck by many participants' comments about the United States and the general assumption that the entire country was rich. I wanted the Appalachian region, an intensely beautiful and mountainous section of the States, to be represented at the 2007 Apimondia in Australia.

From 2005 to 2007, as National Endowment for the Humanities Chair of Appalachian Studies, I did preliminary assessments of pollinator-friendly reclamation methods on surface-mine sites. Many third-world countries are affected by extractive industries such as gold, coal, diamonds, and titanium. Because extractive industries provide quick influxes of money, ordinary citizens and their bees often have to change their agricultural patterns to accommodate them. Although I had known for some time that the Appalachian region is considered a colony by US standards, I was struck by how the rest of the world was not even aware of such pockets of persistent poverty in the States. The Appalachian region supports at least 50 percent of the nation's energy needs by mining its large coal reserves. However, most of its families earn less than $25,000 a year and many require government support to supplement their income. So, seeing Bees for Development succeed in providing specialized extension and education in rural areas such as Africa and India has been a major turning point in my life. If Bradbear and Jackson can do something like start Bees for Development, just like that, and keep the program running successfully, perhaps so could I in my own backyard.

Other European women beekeepers serve the bee industry by showing leadership in associations, the arts (including that of beeswax sculpture), and breakthrough research. While in Ireland and Australia, I met Jacqueline Rocheblave, secretary of Union Nationale de l'Apiculturea; her grandfather had bees in trees, and this similarity provided enough of a basis for our friendship. Her grandfather used stones for lids on his bee trees. He tended to harvest primarily chestnut honey—about one hundred hives at the turn of the twentieth century. Jacqueline's father continued with bees; in fact, she once gave him two Dadant hives to help him increase honey production.

Rocheblave has eighty hives. Her bees also forage on chestnut. She started beekeeping in 1986, leaving Paris, where she had been living, to return to her birthplace in Lozere, Cevernne, in the south of France. Before becoming a full-time commercial beekeeper, she had been a secretary for an American construction firm in which her duties included being a technical writer and evaluating building plans before they were built.

She prefers the bee world, saying, "It is wonderful." As a woman often working alone, she works in stages, taking rest stops in order not to tire. The mountains can make work harder, but she enjoys June, when the hard work pays off in honey sales. In an average year she produces about seven hundred honey kegs; she owns two honey shops.[119]

As secretary of Union Nationale de l'Apiculturea, Rocheblave co-ordinates beekeeper activities, including interest in pesticides, colony collapse disorder research, and economic ventures into meads and varietal honeys. This association has more than forty thousand members, compared to the American Beekeeping Federation in the United States, which generally has fewer than two thousand members. The French leadership regarding colony collapse disorder was one reason why Apimondia met in France in 2009. With Rocheblave's assistance, the association has positioned itself to become a European leader in organic honeys and beekeeping.

In other parts of Europe, the Scots ladies continued a trend of leadership started by Clementine Graham Stirling, whose love of rye skeps was recorded in her translations of literature about them and who preserved a handicraft in skep weaving that continues today. Margaret Logan, assistant lecturer in beekeeping at the North of Scotland College of Agriculture, wrote *Beekeeping for Craft and Hobby* in 1950. This book, a product of World War II mentality, advocated using the "Glen" hive that had fifteen British standard frames. It was, according to Showler, "a heavy hive for heavy flow areas." Showler offers a qualification: "This is a remarkably well-produced book where sadly the authors' comments are addressed to a male audience."[120]

Contemporary Scottish artist Cath Keay uses beeswax in completely different ways from those seen before. She is not a chandler, and she is not a modeler. She is not into cosmetics or waxing down planes. Keay uses bees to create interactive sculptures and in the process studies the parallels between human architecture and bees, with an emphasis on 1930s utopian structures. Working with beekeeper Brian Pool, Keay builds architectural models from sheets of beeswax; "Brian places them in the beehives and takes them out so that they aren't totally absorbed. This process stops me overworking sculpture, making things too precisely finished."[121]

Keay likes to build model buildings from the 1930s because that decade was a time of extraordinary change. According to Juliet Knight, Keay "chose buildings of this era because of her interest in idealism, the utopian view of the future that emerged between the world wars, and the buildings that resulted from it. . . . She enjoys the relationship between the original subject—1930s architecture—and her technique, which uses bees to break down the original model, placing the work inside a bee hive for anytime between one and three months to see what happens."[122]

At times, the bees draw out the cells of beeswax in unpredictable ways, even filling some with honey so that the building collapses in places. Other times, the bees simply nibble at the beeswax. If the season is too cold, the bees hibernate, so Keay's art is subject to the laws of nature. Her interactive art develops by blurring the lines between nature and society. Knight sums up the relationship: "Bees are important to the work in both a symbolic and practical sense: as a highly efficient, structured group of individuals, all working for the collective good, they reflect many of the social ideals of the 1930s. They also help to build the work, or rather to break it down." Keay plans to focus on ceramics and clay in the future.

Some of the most exciting research to emerge on evolutionary bee biology is being conducted by Dr. Gro Amdam in Norway. Amdam's focus is the protein vitellogenin. She did not begin her career wanting to work with bees; she aimed to be a veterinarian. But a series of internships led her to discover that being a vet was not after all what she truly wanted. So, she spent a time "being orphaned," without a goal.[123]

Then, Amdam had the marvelous luck to take a class in population demography. She walked out of the class realizing that she had not understood a thing! The professor, Stig Omholt, used honey bees to show higher-level patterns in fairly simple measures. As someone who was used to learning things easily, Gro was intrigued and challenged not by the bees themselves, but by someone who could teach things that she did not understand. Omholt would serve as her mentor when she decided to apply to the government to work on a PhD in theoretical regulatory biology in 1999. She did not have female mentors in her higher education.

Gro Amdam maintains two laboratories, one in the United States and one in Norway. Courtesy of Arizona State University.

The protein vitellogenin had been studied in the 1970s, but those studies had never really led to conclusions. "All ideas had been silenced," explained Gro. "No one could link the protein to the worker bees in a causal way." So in 2000 Gro began her research using simulation modeling, which predicted that vitellogenin was a factor in nutrition and behavior.[124] The new technology plus the data collected in the 1970s by researchers such as the Brazilian Dr. Zilá L. P. Simões led Gro to another path in experimental biology. (I discuss Simões further in the chapter on South America.) Amdam met Simões in Germany to develop "knock down" methods in which vitellogenin would be downregulated in worker bees. The results from these studies confirmed that vitellogenin affected hormones and behavior.

Amdam finished her dissertation in 2003 and accepted an invitation to visit the University of California–Davis Department of Entomology, led by Dr. Robert Page Jr. After forty-eight hours, Amdam called her spouse to say, "We are moving to California!"

In 2005, after Page began directorship of the School of Life Sciences at Arizona State University, Amdam applied for a faculty position there in US agriculture. Meanwhile, having landed her own research money from Norway, Amdam opened her lab at the ASU School of Life Sciences. Then she began studying vitellogenin's role in the evolutionary framework of social honey bees, specifically reproductive sociality as a predictor of social behavior. She maintains her associate professorship in Norway and finds that the collaborative relationship benefits her research and her lab partners.

Keeping a lab on either side of the Atlantic Ocean has benefits for both Amdam and the countries in which she maintains citizenship. Norway, often called "the last Russian state," does not often promote cutting-edge research for its own reward, yet it will provide generous funding for those who can compete with other participants. The United States rewards its researchers with a high degree of visibility, but it funds only one-tenth of the grants that Norway funds its researchers. So Amdam has been able to publish and build a competitive portfolio in the United States, making her more attractive for Norwegian research funding consideration. She has become embedded in the system of funding, being able to coordinate a US lab with twenty participants and a Norwegian lab with equal potential.

The cross-cultural exchange benefits students, too. The US students receive a first-class international experience in a socialist democracy. The Norwegian students get to experience the United States, where they are expected to perform to higher standards. With these kinds of cultural environments, in which performance is a key part of success and socialization, Gro suggests that her accomplishments neutralize debates about affirmative action. In her case, she explained, the system worked the way it was supposed to. Her education and research opened the necessary doors, without the need for affirmative action safeguards.

No longer defined by expectations regarding patriarchal ideals, contemporary European bee women face challenges as complex as those faced by their forebears. Global warming and genetically modified crops dominate European conversations. Kristin Traynor writes about Danish beekeepers benefiting from global warming: "The Scandinavian countries are the winners with global warming," she suggests, since they have long winters, short days, and cold temperatures.[125] Denmark uses its numerous islands to create breeding stations for its bees. Most of the instrumental insemination is done by men.

Norwegian researchers such as Amdam are using diutinus bees—bees that live through nine months of winter—to study vitellogenin. "The Norwegian diutinus bees store vitellogenin 'like little lunch boxes,'" explains M. E. A. McNeil. Theorizing that vitellogenin may also be an antioxidant, Amdam has been able to show that "when Vg is knocked-down in lab analysis, the bees [are] more susceptible to oxidative stress."[126]

European women beekeepers have been best positioned to act as mediators between the industrialized continents such as North America and the third-world continents such as Africa and, until recently, India. European women lead new types of research that may provide insight into the internal rhythms of the hive and the thermoregulatory patterns stabilizing our pollinators, and may also bring stabilization to conversations regarding honey standards, treatments in the hive, and other challenges regarding contemporary beekeeping.

North America

The Great Experiment, Part 1

Deputy Husbands, True Women, Honey Hunters, and Inventresses

Never was there a country more suited to the cultivation of bees.
—Mary Corré Griffith, "On Bees," 1828

Unlike the continents discussed in previous chapters, North America does not have an ancient honey-hunting tradition. The lack of Native American female bee goddesses stands in stark contrast to the bee-centric stories, songs, or prayers in other cultures. North America once had a native honey bee, until the Miocene Epoch cooled the continent considerably, severely affecting habitat. Although the honey bee went extinct approximately twenty-five million years ago, other pollinators such as solitary and semi-social bees thrived in the ensuing different regions, such as deserts, the Appalachians, the Rockies, and the Great Plains.[1] The European honey bee *Apis mellifera mellifera* (German black) was introduced to North America via human assistance when English colonists arrived in the 1600s. So successful was this entomological introduction that within a few hundred years North America elevated beekeeping to a commercial industry without a central religious figure, dominant language, defining family structure, or political dynasty.

North American women beekeepers have participated in a diverse agriculture regardless of the many religious differences among the continent's Native American, English, French, Spanish, German, and African groups and other immigrant groups of the seventeenth century. Many women from beekeeping countries such as Russia, Italy, Switzerland, Scotland, China, and Poland immigrated to North America in the nineteenth and early twentieth centuries, precisely when technology was making commercial beekeeping possible. North America has benefited from these waves of women beekeepers. Some of the faith-based beekeeping waves have included Dutch Lutherans, African slaves, Quakers, English Shakers, Mormons, Moravians, and Protestants, as well as other utopian religions that have defined North America in the past four centuries. In trying to find one word that would unite them all, North American colonists in the seventeenth century deliberately appropriated the term "bee" to mean a social gathering, such as a quilting bee. It was the first time the word had been used to describe collective human action instead of an insect.[2]

Because of its diverse communities, the United States has offered a smorgasbord of feminine paradigms, some borrowed from Europe and some reflective of emerging North American realities. These paradigms, such as the good wife, the true woman, the new woman, and the nuclear family stay-at-home mother, are just a few of the many patterns by which North American women have been defined.

But these conventions often have been exclusive of minority women such as slaves and Native Americans. They also did not apply to many North American women. In fact, in some seventeenth- and eighteenth-century tribes, Native American women enjoyed much more equitable partnerships within their society than the European transplants did. Since religious rituals respected maternal traditions, women held council seats, served as medicine women, and were subjected to less restrictive marital laws. Children were considered members of the tribe, so child custody was not an issue. Because Native Americans were habitually marginalized from their ancestral lands, they interpreted the honey bee as a symbol of encroachment, calling it the "white man's fly."[3] Although they did become honey hunters,

Native American women in general did not embrace the beekeeping traditions that accompanied more settled, agrarian lifestyles.

Other minority groups did, however. African American women communities use the honey bee image and bee products in their arts, agriculture, and culinary practices. In the past fifty years, Hispanic American women have made contributions to the commercial queen production industry. Asian American women have become noted research scientists and culinary artists. The "great experiment," as George Washington called the fledgling colonies' effort to establish a democracy, eased the way for North American women beekeepers to create an ever-wider set of opportunities. Access to technology, innovation, landownership opportunities, and travel, as well as freedom to create unconventional roles, differentiate North American women beekeepers from those on other continents.

Deputy Husbands

Women came to North America during the seventeenth century for various reasons, but three stand out above all others: marriage opportunities, religious conviction, and widespread European poverty. According to Laurel Thatcher Ulrich, between the years 1650 and 1750, "almost all females who reached the age of maturity married."[4] The Pilgrim women who came during the early part of the seventeenth century were committed to staying in the colonies. Pilgrims were poorer than other Protestant groups such as the Puritans who followed later, but they came prepared with domestic skills. The Puritans were less conservative than the Pilgrims, more affluent, and better educated. They maintained extensive contacts with England, arranging for honey bees to be sent in the 1620s.

Once these established communities began to reduce their high mortality rates, poor English women came to North American colonies as a way to resolve their debts and escape the prisons and poorhouses. The women in these groups brought with them social patterns, such as the good wife, that had to be modified once they arrived in the colonies. Whether women chose to come with a religious group or were forced to come because of economic circumstances, most had been imprinted by the honey bee rhetoric of swarming, hiving off, and worker/drone bee oppositions that was so prevalent in political and religious discourse in all parts of England during the early to mid-seventeenth century.

Scholar Karen Ordahl Kupperman's research suggests that bee imagery inadvertently gave New England women "coping mechanisms" to survive those initial years after arriving at the colonies. For instance, the Massachusetts Bay Colony accepted more than a thousand colonists and thus ensured that the colony would have sufficient workers and resources and a good reserve of morale from which to begin. In this way, the English clergyman Richard Eburne, writing in 1624, was correct in comparing colonies to beehives: "The smallest swarms do seldom prosper, but the greatest never lightly fail."[5]

Since the Plymouth and Massachusetts Bay colonists brought their families with them, the psychological effects of colonization seem to have been tempered in their colonies earlier than in Jamestown, an issue that Kupperman addresses: "The key to getting people to work together on common goals, paradoxically, was to allow them to work for themselves and their families. . . . Families, made possible by the emigration of women, rendered individual efforts meaningful; passing an estate on to one's children was the goal of hard work and deferred gratification."[6] When it was theorized that the colonists would be susceptible to Native American attacks, colonist William Wood refuted that notion. Wood realized that colonists who have property to protect would band together. "When Bees have Honie in their Hives, they will have stings in their tailes," he stated.[7] This combination of theological, financial, and familial interests succeeded exponentially compared to other commercial ventures focused on how much money could be made in America.

Once on North American soil, women had to figure out how to survive with minimal skills and tools necessary for the harsh winters and poor soil. Sewing circles called quilting bees and barn-raising bees provided social outlets and chances to build solidarity. Colonial women in North America had more flexibility with social codes than their European counterparts did. Since men were often away from home because of maritime trade or frontier expansion, women had legal rights to handle money and property affairs in their husbands' absences—in effect being "deputy husbands."[8]

Even though women were not allowed to participate in the political legislation happening in the eighteenth century, their social bees began to take on political importance. When English colonies began to show profits in shipping, timber, slavery, and rum in the eighteenth century, the British increased taxation in order to maintain other colonial enterprises in India and the Caribbean. Suddenly, the hive image became loaded with ideas of political strength. The colonies were enjoying financial and material success. The long distance from Europe and an increasingly collective unit of colonies meant that the swarm image was no longer used to encourage people to leave Britain but to encourage the colonies to leave the empire.

In England, Parliament continued to tax the North American colonies in an effort to bankroll its war on France and extend its empire into Asia. King George III increased taxes on colonial goods, especially molasses and tea. Already angry, the colonists organized against the 1764 Revenue, Stamp, and Sugar Acts as they had at no other political event.

In the resulting turmoil, the honey bee image became an important symbol of rebellion among the colonists. Tax collectors were dubbed "drones," an image that suggested the English were lazy and living off the "worker bee" colonists.[9] No less an authority than Benjamin Franklin encouraged colonists to eschew British products like sugar and molasses and use honey, which was not subject to tax. According to Franklin, the British used forced labor such as chattel slavery in the West Indies to fuel their love of "soft" luxuries such as coffeehouses, tea rituals, and chocolate, all of which depended on refined sugar.

Franklin wrote, "West Indians had the Sugar Cane, from which, by forced labour of Slaves, Sugar and Melasses are extracted, for their Masters Profit. This is denied to us in the Northern Colonies: But then we have an Infinity of Flowers, from which, by the voluntary Labour of Bees, Honey is extracted, for our advantage." Ever the optimistic pragmatist, Franklin even surmised that well-kept hives may encourage "Great profits" and "economic independence."[10]

The anti-British rhetoric worked well on the women in the colonies. Writing in her diary, young Anna Green Winslow captured the excitement and fervor in Boston, which was a social hub for revolutionary action: "Sons of Liberty meeting every day and every night; the Daughters of Liberty holding spinning and weaving bees; the citizens pledging 'to drink no tea till the obnoxious revenue act was repealed.'"[11]

In quieter and less politically charged language, Quaker Elizabeth Sandwith Drinker recorded her daily life in plain, unemotional musings. An upper-middle-class wife, Drinker often managed as a deputy husband when her spouse was conducting business and could not return home. This was especially true during the Revolutionary War. Although she was only two blocks from the Philadelphia wharf, she had a barn with a dairy cow, chickens, goats, and bees. Her diary is a strict listing of events. In her entry dated June 7, 1779, she writes, "Saml. Wharton junr. call'd to ask if we had bees-wax to sell."[12] Jacob Franks also paid the Drinkers a visit to collect a Continental tax. Although the Continental Congress was meeting, Elizabeth did not write of the excitement that others felt; in fact, she was whitewashing rooms. Her diary reflects visits, medical treatments, and the occasional minor irritation with British tax officers.

Yet, the Drinkers were intimately acquainted with Founding Father Dr. Benjamin Rush, and Elizabeth was a good medical practitioner with beeswax and honey—essential defenses against bacteria. Her diary lists trial and error methods with minor ailments. In 1796 she used an ointment made of beeswax, turpentine, and oil called Burgandy pitch plaster to aid someone, noting: "I believe it to be good for a Cough." On April 23, 1799, Drinker wrote: "Dr. Rush talkd of renewing the blister by a plaster of flies—it did not take just as I should like, but may alter for the better—I dress'd it after, cutting with Cabbage leaves, the Doctor order'd a plaster of beeswax and oyl cabbage leaves are out of fashion."[13]

Drinker did not have an opportunity to attend medical school. Her ointments were not unique, but from her diary, we have a better picture of how women used beeswax on a daily basis for minor medical emergencies. A midwife named Martha Moore Ballard used beeswax when tending to a burn on a child named Stephen Jones on April 6, 1792, making an ointment of linseed oil, beeswax, and rosin.[14]

These women's medical efforts highlight just how precarious health care was in North America and how many of the readily available health-care remedies depended on the beehive. The diarists are the voices of medical practitioners who would provide immediate first aid but were denied admission to medical universities and other formalized institutions. Beeswax and honey, specifically the antibacterial properties associated with them, were the best options at hand for a vast assortment of maladies.

After colonial existence, the United States aggressively pursued land expansion as a way to flex its muscles to European powers that had staked claims along the Mississippi River—that is, Spain and France. If that meant encroaching on and breaking treaties with Native Americans, so be it. But this also meant developing expansive agricultural policies that depended on exploitable labor. Blatantly ignoring the mass erosion and soil depletion caused by tobacco and cotton, for instance, were agricultural organizations that used beehives draped by feminine deities associated with the Greco-Roman civilizations to suggest pastoral harmony with the land. These feminine images promoted a fledgling nation confident of its strength and balance.[15]

In other words, just as the Greco-Roman societies had worshipped bee goddesses but marginalized their women citizens, the United States followed the same course. Late eighteenth-century

North America was hardly a model of moderation, and agricultural land policies were anything but progressive. Excess in the form of land use, chattel slavery, immigrant labor, and economic turmoil meant that the fledgling States was as exploitative a society as any that had preceded it. Many women were part of that exploitation, being disenfranchised from owning land and voting. There was one crucial difference: in lieu of a centralized bee deity to serve as a social identity, as Artemis and Diana had functioned elsewhere, the worker bee image was used as a symbol for new American values of thrift, hard work, and frugality.

True Women

As the United States shifted from a frontier economy to capitalism, the paradigm of the true woman replaced that of the good wife. There was less need for the good wife model, in which a woman was expected to be an equal partner with her husband, serve the community, and barter with neighbors. Instead, women were expected to be in the home, providing the primary care for children.

The true woman was a paradigm that nineteenth-century women beekeepers had to negotiate. Even as late as 1906, in an essay by E.J.P., "Beekeeping for Women," the author addresses "the true woman," whom she argues is better fit for queen grafting, comb handling, and

Domesticity during the Nineteenth Century

Similar to the British ideal of the angel of the house, the North American paradigm of the true woman had four characteristics: submissiveness, piousness, virtuousness, and domesticity (Douglas, *Feminization of American Culture*). This paradigm applied to only 10 percent of North American women at most, and it depended on the exploitation of other women, such as black women, Irish or German immigrants, and rural women relocating to cities.

every phase of queen production. "How many beautiful queens are killed in the process of grafting," E.J.P. asks rhetorically, "because they are clumsily or nervously handled by a man who is addicted to tobacco and other nerve-wracking vices."[16] These comments and others like them written far earlier in the nineteenth century provided a steady chorus of encouragement and support for women thinking about becoming beekeepers. Fortunately, although US women would not get to own land until the passage of the Married Women's Property Rights Act in 1848, women did not wait for laws to seek equal opportunities. Religious utopias offered some freedoms not afforded by the Constitution. Among US utopian communities, the Shakers offered women beekeepers equal status.

Arriving from England in 1774, Mother Ann founded the United Society of Believers in Christ's Second Appearing, commonly known as the Shakers for their frenetic celebrations of spirituality. This egalitarian religion emphasized "the Above all," instead of a Holy Trinity, which many Protestant communities embraced. The Shakers believed in a divine duality that embraced both sexes—the male aspect of the Father and the female aspect of Wisdom. They demanded total celibacy of participants. In exchange, the communities offered complete equality between the sexes. The philosophy emphasized that Christ had already arrived on Earth, and thus believers had to adjust their living standards to reflect his impartial love and justice. Using the hive as a metaphor for a productive social unit, the Shakers offered a fundamentally new way to define family in North America.

In order to emphasize equality between the sexes, separate accommodations housed "families," men on one side, women on the other. In buildings built for general use, separate entrance doors also emphasized equality. In these Shaker settlements, women were beekeepers and kept pace with the changes in bee technology just as the male beekeepers did. Unfortunately, we know very little about their daily beekeeping routines since women's bee journals are difficult to find, although men's journals shed some light on Shaker beekeeping in general.[17]

Hancock Shaker women kept honey bees in movable-frame hives. In keeping with the religion's emphasis on equality between the sexes, nineteenth-century Shaker women worked bees just as their male counterparts did. Courtesy of Hancock Shaker Village, #1991-5086.

During the Civil War and Reconstruction, the Shakers used the honey bee in hymns to lift the morale of their members. At the Shaker community in Lebanon, New York, "Busy Bee" was just one of the songs Shaker Anna White "received" in a spiritual vision:

> Like the little busy bee, I'll gather sweets continually
> From the life-giving lovely flowers, which beautify Zion's bowers.
> No idle drone within her hive will ever prosper, ever thrive,
> Then seeds of industry I'll sow, that I may reap wherever I go.[18]

This utopian community did not last through the twentieth century.

In frontier Canada, women used the honey bee as a symbol of community. A woman named Catherine Parr Strickland Traill, who had enjoyed a modest aristocratic but financially strained background, records linguistic usage in her letter dated October 25, 1832: "Now you know that a bee in American language signifies those friendly meetings of neighbors who assemble at your summons to raise the walls of your house, shanty, barn, or any other building: this is termed a 'raising bee.' Then there are logging-bees, husking bees, chopping bees, and quilting bees. The nature of the work to be done gives the name to the bee. In the more populous and long-settled districts, this practice is much discontinued, but it is highly useful, and almost indispensable to new settlers in the remote townships where the price of labor is proportionally high." Strickland herself was the beneficiary of such a bee in 1833, recording that "our neighbors cheerfully obeyed our summons; and though the day was far from favorable, so faithfully did our hive perform its tasks, that by night the outer walls were raised. The work went on merrily with the help of plenty of Canadian nectar, whiskey, the honey that our bees are solaced with."[19]

She records the social class differences in linguistics in August 1833, noting that "the stripping of the corn gives rise among some people, to what they call a husking bee, which, like all other bees, is one of Yankee origin, and is not now so frequently adopted among the more independent or better class of settlers."[20]

Some women were not settlers but established agrarians. Mary Corré Griffith had been involved with social justice issues such as slavery and women's rights before her husband died. Being widowed with eight children changed her destiny: Griffith became a beekeeper and writer to supplement the loss of her husband's income. Throughout her career, she learned that moisture in a hive was a major problem. In 1820 she bought a farm, which she called Charlie's Hope. She corresponded with James Thacher and J. V. C. Smith about bees from the 1820s until the 1830s, and she penned an article on beekeeping published anonymously in the *North American Review*. Griffith invented a hive of her own, the Charlieshope hive, which Thacher praised in his 1829 bee book *A Practical Treatise on the Management of Bees*. She earned the title "inventress" for the hive, its primary asset being moisture control.[21]

Griffith's design was a "T" bar that two hives hung from. This design prevented moisture buildup, since the hives could drain. Rodents could not get in the entrances. From Griffith's notes, it also seems that the hives could be picked up and moved from one location to another. Being influenced by Greek hives, the sides were narrow to give support to the comb. Griffith wrote the details in her anonymous article in the *North American Review:* "The Charlieshope hive is thirteen inches square at the top, and is of the same size at the bottom of the front and back, but the bottom of the sides is only seven inches wide."[22]

Gene Kritsky's book *The Quest for the Perfect Hive: A History of Innovation in Bee Culture* provides a good illustration.[23] From the front, it looks like a conventional Langstroth movable-frame box hive. The back of the hive slants downward, presumably to rid the hive of moisture but also to support the comb. It looks like a Kenyan top-bar hive design. The bottom board was hinged so that the keeper could unhinge as needed. Kritsky provides no comment about its general usage, but others in the region seemed to have adopted the idea of a suspended hive that could be moved from place to place.

Mary Griffith was concerned about the economy of the region and optimistic about the opportunities for beekeeping. She wrote:

> We esteem it a very desirable object to make the care of the bee more common than it has hitherto been, in this part of the country. With the exception of a small one under the superintendence of the society of the Shakers, established in New Lebanon, we neither saw nor could we hear of more than a single apiary, on a journey last summer to Lebanon Springs, although we made many inquiries. Never was there a country more suited to the cultivation of bees. Even in August there is an abundance of white clover, and small springs and shallow rivulets appear at every turn.[24]

Griffith was also a visionary in the literary sense. Many consider her to be one of the first science fiction writers in North America. Writing about Griffith's contribution to American letters and women specifically, Beverly Seaton stated: "Griffith was very concerned with the economic security of women. She had no interest in politics or public life for women, only asking such things as they be allowed to clerk in retail stores and do tailoring. . . . She was both a critic of the business community for its treatment of women and its dishonest practices, and a true believer in the opportunities offered by the American way." Furthermore, Seaton writes: "A woman of strong opinions, Griffith despised dogs, bankers, society doctors, and any man who let women work to support him, such as young men who allow church women to raise scholarships for them to study for the ministry. Women's work, she felt, should benefit women."[25]

Griffith's hive may have improved beekeeping in a regional section of North America, but almost no mention was made of it by the 1850s. Her fiction, which had seemed impossible in the nineteenth century, does not seem far-fetched at all today, although it too has fallen out of favor with audiences.

The beekeeping world needed a hive in which beekeepers could evaluate external and internal conditions. The Langstroth movable-frame hive, invented in 1851, met all those provisions.

LANGSTROTH WOMEN

In the middle of the night, Lorenzo Langstroth, a shy, genial pastor, realized that a relatively simple design change based on the concept of "bee space" could improve beekeeping by enabling a beekeeper to take full advantage of the bees' ability to gather honey and pollen during full honey flows and provide maintenance for them during off seasons.

In effect, bee space embodies nothing more than a 3/8- to 1/2-inch space between suspended and removable frames of wax foundation and from the frames to the walls of the hive. Kim Flottum, editor of *Bee Culture*, defines bee space as the "comfort zone" needed for honey bees to build comb, store honey and pollen, take care of the young, and rest when not working or doing the daily maintenance required of a hive.[26] Langstroth's hive design, published in the second edition of his book *The Hive and the Honey Bee* in 1857, illustrated the principle with brood frames measuring 17 3/8 x 9 1/8 inches high internally. Since these measurements correspond to wine-packing cases, Eva Crane mentions that the original brood box may have been wooden cases from France.[27] With movable frames as an option for examining the colony, beekeepers had more information to manage and provide for a colony of bees.

Before the Langstroth hive, beekeepers knew only external conditions affecting their bees. Swarming was predicted by weather conditions. Beekeepers could not take honey without destroying the hive and bees. Nor could they perform effective maintenance to see if their hives were doing well: beyond weighing the skeps, there was very little they could do. After the movable-frame hive was designed, beekeepers could know internal hive conditions in addition to external ones. The Langstroth hive, along with the smoker, extractor, and comb foundation maker, which followed his invention, propelled beekeeping from a cottage industry to a commercial one.

Women profited from these evolutions in beekeeping. With a

Movable-Frame Hive Terminology

Bee equipment can be a confusing topic. Most beekeepers start with either a package or a nuc. A package consists of a mated queen and worker bees, with no foundation or brood. In good weather, packages can establish themselves very quickly. If the spring season is rainy or there is prolonged cold weather, however, package bees may have great difficulty finding adequate nectar and pollen supplies.

A nuc consists of a mated queen, worker bees, and brood. Nucs cost more, but they can give a hive a better chance of establishing winter honey storage because they start the summer season with more generations of bees.

Generally, the bottom box is called the brood chamber. This is where the queen lays her eggs. In previous centuries, beekeepers preferred to use a taller, ten-frame box, called a deep, for the brood chamber. However, the bees do not mind the size of the box in which they live.

Boxes above the brood chamber are called supers. Supers are where bees expand beyond the brood chamber, drawing out more beeswax foundation for the queen to lay more eggs and the workers to store honey and pollen. For women and people with back problems, I recommend using all shallow or medium supers (no hive that was easily manipulated, women could be among the ranks of honey producers and queen rearers and take leadership roles in bee associations. Since the nineteenth century, beekeeping has become an industry that can involve family members and be adjusted to the demands of rearing or nursing infants, teaching skills to children, or employing extended family.

The matriarchal figures in Langstroth's life provided the courage, economic security, and emotional stamina that would accompany his travails as the Father of Modern Beekeeping. An early prominent influence was Langstroth's grandmother Elizabeth Lorrain Dunn. She had been married to James Brown Dunn, a Maryland politician and slaveholder. After her husband died, she was persuaded by the Methodist theology that slavery was wrong. She became a staunch abolitionist and freed her slaves, although it meant losing

deeps) and/or eight-frame supers instead of ten-frame equipment. When I first started keeping bees, I thought bees had to be kept in a deep and a super. Now, I keep bees in medium supers because they weigh less and are easier to work when the colony is full strength in the summer. Since I do not like to use deeps, I make sure that my bees have three medium supers (two supers of brood, one super of honey) before winter approaches.

A wire frame is called a queen excluder, and it does exactly what it says: it keeps the queen confined to the brood chamber. For first-year hives, I do not use queen excluders because I want the bees to build up winter honey stores. For second-year and older hives, I generally put queen excluders on in the middle of June in my region. As winter approaches, the bees will downsize and then cluster to the size of a basketball, so two mediums is plenty of room, with the top super being a honey (or "feeder") super.

The hive needs to have a screened bottom board, which is a passive form of varroa mite control. The hive also needs to have an inner cover, so that bees do not caulk the cover to the top of the super, and an outer cover to shelter the hive inhabitants. Bees technically do not need a bottom board. But bottom boards make life much easier for beekeepers when bees need to be moved or worked. Especially since varroa mites have appeared, screened bottom boards are an important asset.

Anna Langstroth, Ohio, United States. "I won my wife by mathematics," Lorenzo once wrote about his wife. They were married thirty-six years. Courtesy of *Bee Culture Magazine.*

a good portion of her money. Given her French Huguenot heritage, she was also able to impart enough of the language that Langstroth could converse with his beekeeper friend French immigrant Charles Dadant, who was bilingual, and Dadant's French-speaking wife, Gabriella.

Langstroth married Anna Tucker, daughter of Mrs. Harriet Tucker, administrator of a young ladies' school in New Haven, Connecticut. Langstroth taught in Tucker's school once he had finished his divinity training at Yale. Langstroth had an amazing reputation as a kind and charismatic teacher, and those skills seem to have helped in his courtship of Anna. In an article titled "How He Became Interested in Bees," Langstroth wrote about the early years of his marriage:

> At the time of my marriage, my mother and one sister became entirely dependent on me, and made their home with us as boarders in the family of one of our deacons. When I began housekeeping, in the spring of '37, the inflation of prices in the time of Pres. Van Buren had culminated. I paid 15.00 for my first barrel of flour; and although my salary was considered a good one, and was paid promptly every quarter, it soon became quite apparent that my expenses would exceed my income.

> My dear wife, instead of even intimating that it was hard for us to
> begin the world with expenses much greater than would suffice for a
> considerable family, always encouraged me in doing my duty for the
> relief of the dear ones whom God had made dependent on us, saying
> that we might thus safely trust events to our heavenly Father.[28]

With two children, one of whom was afflicted with a spinal disorder,
Langstroth started reading about honey bees. In 1848 Langstroth
established a school for young ladies in a building with a second-
story piazza and a number of spare attic rooms. Using the rooftop,
Langstroth created a small apiary in Philadelphia.

Leaving Anna in Philadelphia, Lorenzo went to Greenfield, Massachusetts, to live with his sister Margaretta; he stayed for six years.
Anna found a place as a teacher in a girls' school. Lorenzo wrote in
an article:

> When I parted from my beloved wife, in Nov. 1852, the future was
> mercifully hidden from me. I was to have no settled home for nearly
> six years, and for more than three-fourths of that time I was to be
> separated from dear ones! I had recovered from my head troubles,
> and at once began to write my book on bees. The larger part of the
> manuscript was sent, as fast as written, to my wife, to be copied into a
> legible hand for the printers; for in the ardor of composition I wrote a
> scrawl which only she and I could decipher.[29]

When Langstroth figured out the concept of bee space, beekeepers
took advantage of it remarkably quickly. His efforts to patent the concept took so long that commercial bee suppliers were already marketing the Langstroth hive before the patent was complete. So Langstroth
would not profit financially from his discovery. He continued to have
major health breakdowns, and with economic recessions following
the Civil War, he and Anna remained in poverty.

But Langstroth did write a very successful book, *The Hive and the
Honey Bee*, which has been reproduced multiple times. In this book,
he emphasizes that women, like the honey bees, should have access

to the fresh air that beekeeping provides: "In the very land where they are treated with such merited deference and respect, often no provision is made to furnish them with that first element of health, cheerfulness and beauty, heaven's pure, fresh air."[30]

Anna Langstroth died in 1873. In writing about her to the Dadants ten years later, Langstroth tried to explain her contribution:

> Friend Dadant, My dear wife died in Jan'y 1873. In carrying on my large correspondence and in many other ways she was as much to me in my seasons of prostration, as Huber's wife was to him in his blindness. Without her labors, out of school hours, that winter, it would have been impossible for me to prepare my work for the press. Never once did she even intimate that she was overtasked, and all her letters breathed such an unselfish spirit as can be attained by only the loftiest and purest characters. I shall say no more, at this time, of this beloved companion than to put on record the fact that, in our married life of over thirty-six years, I cannot recall a single experience in which I knew her to seek her own happiness at the expense of others.[31]

As a postscript, contemporary Langstroth scholar Marc Hoffman's note is a good summation of the importance of Anna Langstroth: "She is always described as being a mere copyist, making LLL's handwriting legible. But we know from her letters and from her formal writing, e.g., her plea to the patent office for the 1863 reissue of the patent, that she was a fine writer."[32]

With Anna's support, Lorenzo Langstroth shaped a world that was more advantageous for beekeepers. Most of the women discussed from this point forward were influenced by the Langstroth movable-frame hive. This hive was taken to all areas of US expansion, Australia, and South America. It enabled women to have more control of their bees and, if needed, opportunities for supplemental finances for their families. Because the Civil War exacted an enormous toll in lives, many women became beekeepers, and the movable-frame hive undoubtedly was their hive of choice.

TEXAS DIANAS

The elusive huntress Diana was the Greek goddess known for her beekeeping and hunting prowess, so she was an appropriate symbol for Texas frontier women who became beekeepers. Texan bee historian Clark Griffith Dumas used the phrase "Texas Diana" to describe nineteenth-century bee ladies who benefited from the ample floral opportunities afforded by the Plains and the western-state prairies.[33] After the Civil War, many women were left as widows or had husbands who were maimed or wounded. The financial duties fell to them. Mrs. Sarah Elizabeth Sherman, Mrs. Jennie Atchley, and Mrs. Florence Weaver were the most visibly successful commercial beekeepers in Texas.

Widowed in 1868, Sarah Elizabeth Sherman moved to Salado, Texas, in 1870 to become one of Texas's first foremost commercial honey producers. She consistently harvested six thousand pounds of extracted honey from sixty hives in four apiaries. She also wrote for *Texas Farm and Ranch* magazine on a regular basis.[34]

Moving to Beeville from Dallas in 1893, Jennie Atchley, a queen producer, appeared in *Texas Farm and Ranch*, which highlighted her "800 and 1,000 colonies of bee devoted exclusively to queen rearing." Some of her queens fetched $100 in the 1890s. Atchley also wrote articles about bee plants, which were necessary for her queen nucs to succeed. Describing each quadrant, she wrote, "You can look over the territory I have described, and suit yourself in location, and you need not be afraid but you will get some honey anywhere in Texas, but out in the hills of Southwest Texas is the best place."[35]

Jennie had ten children, "five boys and five girls—and all have pleasant employment at the home place."[36] She managed to publish the *Southland Queen,* a bee magazine. It merged with *Lone Star Apiarist* in 1902 and ran for only five issues until 1904, when the family moved to California.

Of the Texas Dianas, the woman who left a remaining legacy was Florence Weaver, whose family still shapes Texas beekeeping. The first of eight children, Florence Somerford brought ten beehives as her wedding dowry to her marriage with Zachariah Weaver in Navasota, Texas. Her brother Walter gave them to her, never imagining that the Weavers would establish an international bee business that lasted for five generations. In fact, even as children, Walter and Florence shared a mutual passion for bees, combining their resources to order A. I. Root's *ABC and XYZ of Bee Culture* and subscribe to the magazine then called *Gleanings in Bee Culture*. No doubt Florence needed the honey and beeswax because, since her mother had died, she had assumed responsibility for her siblings' upbringing.[37]

Zachariah saw the potential his wife's hobby could bring to the family. While Walter went to Cuba to make his fortune in honey, Florence and Zachariah stayed in Navasota and reared queens and children. Florence took responsibility for thirty-two children, only eight of them her own. Thus, when asked for a word to describe his grandmother, grandson Binford Weaver exclaimed, "Indomitable."

Honey Hunters

DOLLY WEBSTER: CAPTIVE OF THE "COMANCHE SAVAGE WILD"

A unique literary genre in North America is the captivity account, in which Native American Indians take English colonists or settlers hostage. Depending on the circumstances, the captives are exchanged for food or weapons or Indian captives. Sometimes, in cases of drought of warfare, the hostages are killed. At times, the hostages become so accustomed to Native American ways that they are adopted by the tribe. The plot of most captivity narratives includes a Christian intervention moment, such as an angel or conversion experience. Because English expansion defied the boundaries of its treaties with Native Americans on a regular basis, captivities and ensuing narratives appeared well into the nineteenth century.[38]

A few captivity accounts detail how Native American women adapted to honey bees on the North American Plains. Dolly Webster was moving to the Texas frontier when she had the misfortune to be taken captive, first by Comanche Indians and then by Caddo Indians. While her writing contains some characteristics that define Christian captivity narratives, it remains a fine example of how humans worked like honey bees in terms of delivering information. Captain Kerns had learned of her whereabouts from a Mexican prisoner he had taken hostage after a skirmish with Caddo Indians. It is fortunate that Kerns knew of the Comanche movements, for the Tywackness Indians, who were cannibals, had begun negotiations to purchase Dolly and her two children, Patsy and Booker: "They [the Tywackness] said whites were the best eating in the world."[39] It is impossible to tell from the narrative whether Dolly was being teased or if there was indeed some truth that white people are more succulent than other races when served as a main course.

The Caddo chief refused to trade Dolly and her daughter but did accept a mule for her son, Booker. In a poignant turn of the narrative style, Dolly inserted a poem to describe that loss:

> O! these dread walks, those hideous haunts of pain;
> I've traced new regions, o'er the pathless plain.
> And gave to the Comanche savage wild,
> My only boy, my tender darling child.

Grief aside, Webster wrote a fine travelogue of her travels with the Comanche and Caddo Indians. Although she deplored the Comanche, she found the Caddo very friendly and hospitable. Even after she had been returned to the Comanche, the Caddo chief would look for her at meetings and express concern.

Dolly described the beautiful landscape by the Rio Faco: "The valley is environed by high hills on each side nearly perpendicular. This valley is about two miles long, by one in breadth it is very rich and well timbered; there is also a great variety of springs breaking out of the mountains on all sides—it is from these springs that the Rio Faco has its source. This valley is one of the most beautiful natural curiosities I ever saw. The water from these springs is disagreeably warm."[40]

She also wrote about the honey hunts that women participated in as well as men. By now, she and the Comanche had traveled to the San Saba River. While the Indian men from various tribes organized a raid on white territories, the women procured honey from the caves. In this, the Comanche used their white captives, and thus Dolly found herself with a rope around her, "let down to a place where the bees were flying in all direction":

> Alas, all my efforts failed in procuring the honey, being suspended by my arms, upon which my whole weight hung, over a precipice of more than one hundred feet high; thus situated and thus confined my exertions proved unavailing, in consequence of which, the squaws beat me cruelly, and while thus incensed, they burned nearly all my hair off my head, while my hands were badly burned. The bees were in a ledge of rocks, perhaps 100 feet high or more, . . . and while thus suspended by the rope, and undergoing the torture of whip and fire, I suffered all the horrors of death, for head and hands were literally burned into crisp. . . . They procured no honey from this place, but obtained an abundance from the trees.[41]

Perhaps not surprisingly, Dolly began to plan her escape after this incident, but her son Booker, with whom she had been reunited given the extensive trading among the Indians, refused to go with her.

Dolly was not the only person to record Native American women and honey hunting. Jimmie Hart, a young Irish immigrant who settled in the San Patricio region of Texas, was eleven when he was captured and taken to San Saba County. Jimmie refused to eat raw horse meat and thus was starving to death. "One day," however, "an old squaw got an ax and taking the boy by the hand, led him to the woods. He was helpless to defend himself, and thought the worst had come, but instead, this old Indian woman knew where there was a bee tree in the woods nearby, and was taking the ax to chop into the tree to get the honey. The boy ate this pure food with genuine relish, and later in life related how this honey saved his life and put him on the mend.[42]

168

Inventresses

ELLEN TUPPER: EDITOR, BEEKEEPER, QUEEN PRODUCER

Ellen Tupper is a fascinating figure in the American beekeeping industry, not just because of her contributions to bee extension on the Iowa frontier, but because in her efforts to provide for her family, she entered a moral gray area where few women beekeepers have been documented. Tupper is a good case study to show how in an age that assumed women to be morally superior to men, women could make poor judgment calls.

Tupper was descended from a well-respected New England family that landed with the *Mayflower* and then settled in Rhode Island. Her mother, Hannah Draper Wheaton, died when Ellen was young, leaving Ellen to raise her brothers and sisters. Her father, Noah Smith, moved the family to Calais, Maine. She met her husband, Allen Tupper, a prosperous lumberman, and they married in 1843.[43]

The Tupper family's move in 1851 to Brighton, Iowa, was because of Ellen's precarious health. At this point, the family's fortunes, which had seemed so bright with their histories, ambitions, and educations, took a turn for more difficult fates. Allen struggled to make the transition from a well-established mariner town to the frontier. Specifically, Allen had wanted to become a minister, according to a family history written by Jere True and Victoria Tupper Kirby, "but an attack of fever put an end to that dream."[44]

Ellen Tupper soldiered on. When teaching in the late 1850s, she rode to school on a horse, one child sitting in her lap, another sitting behind her. She faced other hardships typical of women on the frontier. One of her children died of cholera when an epidemic swept through Brighton. Tupper started keeping bees in 1860. A hurricane destroyed her apiary one year; during a harsh winter, a fire destroyed her home and hives.[45]

In spite of the frontier challenges, Iowa had cultural freedoms

Ellen Tupper, Iowa, United States.
Courtesy of *Bee Culture Magazine.*

that benefited the Tupper family. Ellen taught beekeeping at Iowa Agricultural School. Iowa also had become a bastion of Christian socialism. Between the influence of Iowa State College's equitable admissions and the religious movements in the region, Ellen's oldest daughter, Eliza, converted to the Universalist Church and then later followed Unitarian theology. As a pastor, Eliza established at least ten schools in Minnesota and the Dakotas. There is no indication that she followed in her mother's bee boots, but Mila, a younger daughter, did, rearing bees and becoming a Unitarian minister in Minnesota before living in the Dakotas. Another daughter, Kate Tupper Gilpin, became a college professor at the University of Nevada. A son, Homer, stayed in Iowa. Margaret, the youngest, married a wealthy Texas business-man and provided Ellen shelter during the last years of her life.

Throughout her career, Tupper contributed to the Iowa State Ag-ricultural Society reports, *American Bee Journal,* and the *Prairie Farmer.* When writing an annual report on the state of beekeeping in Iowa, Tupper argued for more stringent codes of judging at the state fairs: "When a premium is offered for the best specimen of honey exhib-ited, a man who has but one stand of bees, and hardly knows a drone from a working bee, may chance to have for that time the best box. . . . It therefore seems as if the bounty should be offered to the one who obtains the most honey in the season from a stand of bees or to the one who has safely increased the fastest, or for the best average yield from a certain number."[46]

When Charles Dadant needed financing to buy Italian queen bees, Tupper provided the money. She was invited to speak at a three-day event in Washington, Iowa, in January 1871. Apparently, the audience was so absorbed by her that "the Institute forgot dinner, and adjournment was not taken until 1:00 P.M."[47] In 1872 she moved to Des Moines to start the Italian Bee Company with Annie Savery. When *American Bee Journal* needed an editor, she joined the staff in 1873, staying until 1875. By all accounts, Tupper seemed to be in control of finances, family, and future endeavors. Surely, one would think, the vagaries of frontier life were over.

In 1876 Tupper's life tumbled precipitously. Tupper was supposed to coordinate the 1876 Philadelphia Centennial bee exhibit, which should have been a rewarding task for a well-respected New Englander. It would have been the perfect blending of her family's history and her passion for bees. As late as February 1876 she was soliciting exhibits in *American Bee Journal*.

Yet, on January 28, 1876, Tupper was arrested on forgery charges. Even as she was starting a queen production business, editing the *American Bee Journal,* and managing a family, Tupper had been forging checks to cover financial shortfalls. "The Queen Bee's Temptation" screamed the *American Bee Journal* headline in March. Editor Thomas Newman explained in a brief paragraph: "It appears that she freely used the names of relatives and friends, and in addition, forged the names of leading citizens of various cities of Iowa, from the name of the governor of the State, down; as well as the names of leading men in the East."[48] Tupper even signed a few checks "Jesus Christ."[49]

Editor A. I. Root was much quieter in *Bee Culture,* which was odd, given that he ran a column titled "Humbugs and Swindlers" and could be merciless when he was riled. Instead, on the front page of the March issue, he merely asked that his readers consider kindness: "Mrs. Tupper is to be pitied as everyone who is taken in by temptation step by step."[50]

Evidence of a breakdown was occurring as early as a year before Tupper's arrest. Charles Dadant, assuming that Tupper was withholding travel funds from him because she was cheap, angrily wrote in *American Bee Journal*, "In France, we have a saying *souivent femme varie*, often women vary, and Mrs. Tupper shows she is not an exception."[51]

Tupper responded, saying that "a woman who has followed beekeeping persistently for seventeen years can hardly be called 'fickle.'" Intriguingly, she wrote enigmatically, "Leave it to them also to say, if after having spent over three thousand dollars in the one branch of importing, she is not very wise now to let others bear the expense of further risk in the matter."[52] Was this the reason for Tupper's sudden downfall?

Root himself may have provided another answer in a brief paragraph in his "Honor Roll." Even before Tupper's fall from grace and imprisonment, Root wrote about his concern for her: "Mrs. Tupper's life has been, and probably will be one full of active work, many times it seems more laborious and full of business cares than one of her sex ought to bear. Her health of late has been poor and we trust her friends en masse would be glad to see her take more rest and enjoy her bees more in peace and quietness undisturbed by busy traffic.[53]

From a much broader perspective, another reason may have had to do with the irregularities of operating a queen production business with an erratic postal service. Many customers would complain that upon arrival, the queen bees were dead. Several queen producers felt that customers were taking advantage of them, using the postal service as an excuse for the delivery of dead queens. Fake postmaster claims were routine, and some queen producers were swindled. Mr. William King from Kentucky wrote an especially impassioned letter in his defense after he had been included in the "Humbugs and Swindlers" column:

Very many of my orders I filled the fourth and fifth time, complaints coming that the queens died in the cage before being released. . . . Several sent what purported to be statements of Post Masters and Express Agents, certifying that they were dead or dying. . . . All such complaints were listened to attentively, and more and more queens were sent. I became suspicious that I was being played off on . . . and it turned out that some of these so called certificates of P.M.'s and express agents purporting to be signed by the same parties were in fact written in different hands. I feel fully satisfied that I was swindled out of at least 200 pure queens.[54]

The vagaries of Iowa weather were also a factor in the early days of bee business. Railroad companies hated shipping honey bees, and often these bee shipments were given the worst service of any livestock commodity. So it is possible that Italian queens did indeed die en route to customers because of the poor quality of transportation.

Ultimately, Tupper, who by the 1870s was a nationally known personality, stopped completing orders and started signing bad checks. Piecing together her movements from bee journals, national newspapers, and her family's records, it seems Tupper went to the Dakota Territory.[55] The Davenport, Iowa, sheriff had to bring Tupper to Iowa to face charges. The April 1876 *American Bee Journal* reported that Tupper was committed to the Iowa Insane Asylum. Letters poured in from readers across the nation, both asking about her health and confessing to being wronged in business dealings. Newman, then editor, made a pointed rejoinder to those who claimed to have paid Tupper and didn't want to pay *him* again for their subscriptions, "forgetting that two wrongs will not make one right." Newman goes on to write, "Evidence accumulates every day to prove that she has been recklessly carrying on this crookedness for years."[56]

Newman's stance was supported by a *New York Times* article of February 5, 1876, "Forgeries by a Woman." The writer states that forged checks started appearing in 1875 but some cases were handled quietly. The writer attributes the cause to Tupper's being "much embarrassed in money matters." Tupper had two daughters in school, and the writer acknowledged that Allen was an invalid for much of the time. A breakdown appears to have happened, although Ellen never mentioned such in the journal she left with Margaret, her youngest daughter. The *New York Times* writer then explains: "As soon as she resumed labor she appears to have resorted to this means to relieve her financial pressure as often as necessity required redeeming her fraudulent paper with money got by either forged paper with money or replacing it with fraudulent drafts."[57] While Reverend J. E. Lockwood would write regularly to defend Tupper's business transactions, others railed with heavy rhetoric against her. Just listen to Mr. Ellsworth: "The unfortunate lady, whose mental and moral machinery has no balance wheel, has fleeced me of 80.00."[58]

Newman never passed up a chance to publicize Tupper's troubles. Tupper's fall from grace was reported regularly in *American Bee Journal.* For instance, an article in the October 1876 issue provides more details about how Tupper was arrested. Apparently, she was indicted for debts of $13,000 and forgeries on notes. She was held in jail until her trial.[59]

In 1877 Newman reported, "Mrs. Tupper was tried for forgery in Davenport, and upon the plea of insanity, she was acquitted and is now in Dakota on a farm." Fairly straightforward reporting until one reads the concluding sentence: "The 'insanity dodge' is quite an institution for all kinds of misdemeanors nowadays and gets 'many a one' out of trouble."[60]

A women's historian named Phebe Hanaford, Tupper's contemporary, acknowledged Tupper's fall from social standing: "[Tupper] broke down in both body and mind from overwork, and passed through a long time of sickness and nervous prostration. After this sickness, her mind appeared to be affected, leading to embarrassments in business, so that at present, she is not active in her chosen sphere."[61]

In 1888 Tupper visited her youngest daughter, Margaret Tupper True, in El Paso, Texas. While there, Tupper died suddenly of heart trouble. Although Thomas Newman's obituary railed, "Frailty, thy name is woman," other people were more demure in their judgments.[62] For all the turmoil of her life, Tupper was written about by her peers as one "democratic in spirit": "Indeed, it would be difficult to find one who had more absolutely escaped the consciousness of social lines," Frances Willard and Mary Livermore wrote diplomatically in their landmark work on important American women.[63]

Undoubtedly, for at least ten years, Tupper influenced beekeepers migrating or immigrating to the expanding frontier and needing beekeeping instruction. Her unassuming manners erased the pretensions that divided people from one another. She imported Italian honey queens, thus providing access to one of the most important strains of bees to transform large-scale agriculture. Just as important, she taught her children to promote justice and rights for women in the West. Her children became ministers, teachers, and professors.

But Tupper also serves to place in context the difficulties facing women beekeepers in the nineteenth century. Many ideals—some realistic, some not—were placed on nineteenth-century women, mothers in particular, the most persistent of which is that women were supposed to be more moral then men. The greatest difficulties for queen producers, such as maintaining cash flow, affect contemporary beekeepers as surely as they did those of the nineteenth century. If the path to misery is paved with good intentions, Tupper is a good example that starting a queen production business with poor preparation could lead one down that road just as easily as any other career.

Fortunately, North America is imprinted by the idea of an "Eternal Frontier," to quote Frederick Jackson Turner. This idea is superimposed on everyone, including Ellen Tupper. North America is big enough, and thus often forgiving enough, to allow a person a second chance. Tupper acknowledged her mistakes and moved West, where her children were establishing careers for themselves. In the new world made possible by movable-frame hives, Tupper made unwise choices in a risky and emerging new industry, but the bets she placed on her children paid huge dividends.

FRONTIER BEEKEEPERS

In other parts of the frontier, women beekeepers did well to participate in their local associations and attend conferences. In a pamphlet printed by the National Beekeepers' Association, Colorado, in 1902, nineteenth-century women were honored as successful beekeepers: "Before the introduction of alfalfa into Colorado, wild flowers furnished scanty supply of nectar, and often bees died of starvation. With alfalfa came the red, white, alsike, and sweet clovers, until now, thousands upon thousands of acres of alfalfa are lined with sweet clovers. Mr. Lykins neglected his bees, but then after he married, his wife took charge of the bees, and bought twenty-five colonies of Italians from Mrs. Baker of Upper St. Vrain, paying eight dollars a colony for them. She sold honey to the amount of $175.00 the first year, 1885."[64]

One Wisconsin beekeeper, Mrs. Frances A. Dunham, produced wax foundation for sale and served as treasurer of a committee to honor Lorenzo Langstroth. Thomas Newman praised the foundation that Dunham produced: "The base of the cells being very thin and the side-walls heavy, [it] supplies plenty of wax to complete the full comb."[65] He even ran an advertisement promoting her foundation.

California quickly became an industrial countryside, in no small part because a beekeeping industry developed along with large-scale specialized agriculture. In 1876 a letter from Samuel and Elizabeth Marshall detailed the successful year that they were having with their bees. The first part of the letter is written by Samuel: "We think we will have over 200 swarms. Our bees are a cross between the common black bee and the Italian bee, which makes for better workers than either pure Italian or pure black bees."

Since the Marshalls missed the mailman, the letter sat for a week, giving Elizabeth a chance to add some details. The Marshalls encouraged their family to come to California because the local Indians were being moved to reservations in the following year. Thus, land would be cheap and, at least for that year, pleasant: "We are having new potatoes today with plenty of butter, honey, and mutton, and we have apple trees, peach trees, olive trees, almond trees, pear trees, and figs all growing nicely. Four years ago we commenced beekeeping without a dime and nothing but my hands and a willing mind . . . and now I [Samuel] have a place worth 2000 dollars and I have a good range that will give me more than 2000 dollars worth of honey every year."

For all their successes, Elizabeth admits that frequently "they were as strangers in a strange land." The closest church affiliation was Adventists, although occasionally a Methodist preacher could be found. Nevertheless, Elizabeth states clearly that "we find God in the mountains, in the valleys, in every thing. This has been a good winter, with nice rains but never mud, the fields green and the ground covered with flowers of every size and color from the tiniest star to the gay wild bee that climbs to the top of the sage, some of them ten feet high."[66]

Elizabeth Marshall was not alone. W. J. Whitney wrote in *American Bee Journal*, "It is true that many of the bee men are living without women, 'baching it,' but many more are not. The majority of settlers here have wives and families and more would have if they could get them *worth having*."[67]

BEE JOURNALS

North American women wrote frequently to the two longest-running bee journals: *American Bee Journal* and *Gleanings in Bee Culture,* or *Bee Culture.* The editorial boards of both journals sought to publish a wide variety of voices in order to learn more about beekeeping, since many readers could not go to school and much of the training was simply word of mouth. The two journals could not have been more different or, fortunately, more complementary.

American Bee Journal started under the leadership of Samuel Wagner in January 1861, but publication was suspended during the Civil War. When publication resumed in 1866, the editorship changed several times, including Wagner, Wagner's son, Reverend W. F. Clarke, and Thomas Newman in Chicago. Newman later hired George York and Ellen Tupper as editors for a short time. *American Bee Journal* eventually catered to an international readership once the Dadant family took the helm in 1912.

Editor of *Bee Culture* A. I. Root, a nineteenth-century self-made man, found in beekeeping the ideals he sought in other areas of life. He was the compiler of *The ABC and XYZ of Bee Culture* and the owner of a beekeeping and candle supply company. A pragmatist to a fault, he supported women beekeepers and hired women workers in his factories and press rooms.

Both editorial boards published global news in an effort to learn beekeeping experiences from their German, British, Italian, French, and African or Caribbean Island counterparts. Women appreciated these efforts. In Root's "Honor Roll," Annie Mann from Ontario wrote in to say that she was "so pleased" a writer named "P.G." was a lady: "I do not know of one lady in Canada who keeps bees, tho there may be." P.G. turned out to be a partner in the A. I. Root firm.[68]

In the columns devoted to correspondence, both journals printed letters from women expressing anxiety, pride, sadness, and glee in the company they found in the pages. For many, writing was a way of finding their identities beyond their communities, and these women's letters record regret, happiness, loneliness, independence, and intelligence as more women became skilled keepers.

Even marital dirty laundry was aired in the bee journals, reflecting how women's divorces could affect their professional reputations. For instance, in the March 1876 issue of *American Bee Journal* there is a brief note indicating that Mrs. Spaid's Honey House in New York had closed.[69] At first glance, this closing seems the result of an unfortunate turn of the economic tide. But Mrs. Spaid had been married to Mr. Perrine, a man who first tried migratory beekeeping on the Mississippi River. In an argument about honey adulteration, a particularly vociferous if not always accurate writer, M. M. Baldridge, accused Mrs. Spaid of slandering Mr. Perrine's honey in order to drive up a market for New York honey. According to Baldridge, Mrs. Spaid had become jealous of Mr. Perrine's fortune.[70]

Was Mrs. Spaid adulterating her honey? Was this why she closed her business? Or had people shunned her business once she remarried? We will never know. To his credit, Perrine wrote to Baldridge clarifying the cordial situation between him and Mrs. Spaid. Baldridge wrote a letter titled "Amende Honorable—Errata" in *American Bee Journal* softening his stance.[71] Intriguingly, Mrs. Spaid never wrote to defend herself or explain why the business closed.

Motherhood was constantly addressed in the journals. Many women were concerned about the difficulties of balancing beekeeping with motherhood. Writing in 1885, Mrs. L. M. Crockett notes that "women that left the care of her family for some other employment stepped down—for what higher holier calling can a woman follow than caring for her family?" She goes on to give excellent advice in the sense that she backs away from the either/or choice implied in her initial question. Instead, she suggests that if a husband is interested in beekeeping, a woman should interest the entire family in beekeeping.[72]

Leadership was also proudly announced. An especially feisty speaker named Mrs. J. N. Heater of Columbus, Nebraska, presented a paper titled "Woman as a Bee-Keeper" at the 1893 Nebraska Beekeepers Meeting. She began by stating, "I never did feel willing to grant to the lords of creation exclusive rights to anything," and she concluded, "These are but a few thoughts. . . . We now leave it to the wiser ones to suggest further why we should or should not enter into this field of labor, to possess it."[73] Mrs. Heater managed an apiary of 150 colonies. She was the cover girl of the September 1895 issue of *American Bee Journal*.

My favorite beekeeper is Lucinda Harrison, who, when crowing about her colleague becoming president of a bee association, weighed in proudly: "Hit him again! Knock the chip off his shoulder! Mrs. Cassandra Robbins of Indianapolis, Ind. was voted President of the Indiana State Beekeepers Society."[74] Other women ended up being announced as presidents, but Harrison's strong, assertive voice still resonates for this researcher.

Lucinda Harrison, Illinois, United States, commercial beekeeper in the Plains. "I confess to a weakness for wanting my own way, and I generally manage to get it as far as the bees are concerned." Courtesy of *Bee Culture Magazine*.

The practical aspects of beekeeping were discussed in the journals' pages. When a colony had turned into robbing bees, Harrison recommended stuffing the entrances with cloths saturated with oil kerosene: "It is amusing to see how soon the marauders are converted into law-abiding subjects."[75] Harrison also suggested that "the man who will keep his bees in old dirty rotten hives deserves not only to have them in his hives but also in his coffin."[76] Regarding swarms, Harrison recommends, "eternal vigilance is the key-note of success in beekeeping." Part of the business of swarm prevention is "pulling hives to pieces," she writes, and to have queens on hand, "save surplus queens from the best colonies, and then they will be ready for use if any vacancies occur."[77]

The business aspects were duly noted. As previously mentioned, Mrs. Frances Dunham advertised her version of comb foundation. Lizzie Cotton advertised her book and queens, and in the days before classified ads, some even advertised for jobs. Mrs. A. S. Keyes wrote to *American Bee Journal* that anyone who was "interested in enlisting the aid of a young woman as a bookkeeper and typist in exchange for tutorials in the bee business could correspond" with her.[78]

Appeals to help fellow beekeepers were solicited, a tradition that still continues. Lorenzo Langstroth often accepted the generosity of the beekeepers. His daughter Anna Cowan wrote letters to a committee treasurer, Mrs. Frances Dunham, and the last years of Langstroth's life were easier because of this financial assistance.[79]

Disgruntled women aired their complaints and frustrations. In one letter to *American Bee Journal,* the writer complained about the discourtesy shown to female association members at the Chicago convention: "I came to learn all that I could, and put up at the hotel to converse with beekeepers, but I was expected to go up into the parlor, and the men remained below in the 'office' where it would not have been considered proper for me to remain. . . . I do not want any favors in a bee meeting on account of sex. But there were ladies present who cared nothing about apiculture and they came with their husbands to see the city." The fact that those same ladies could vote as members was objectionable to the writer. She concluded with a plea: "There is a great deal said now about employment for women, and they will be crowding the ranks of beekeepers, and do let them have a chance."[80]

Newman, as editor, responded in kind: "Some of the matters complained of can only be corrected by the advancing sentiment of the age, such as hotel etiquette, etc." He also agreed with her suggestion about differentiating between the ladies who were apiarists and those who were visitors.

But for the most part, the journals were a good place to advocate for the benefits of attending a convention. Lucinda Harrison urged her fellow sisters "pining at home" to attend the International Convention in New Orleans on February 24–26, 1885. She offered the pragmatic consideration of low fares and suggested that "the cost would be no more than that of a new dress, which a woman of ingenuity could do without by renovating her old ones." The benefit: "A dress would soon be worn out but the rich food for thought, gleaned at the Exposition, would be fresh and bright throughout life."[81]

The Linswick sisters—Cyula and Nellie—wrote to bee journals frequently, and their letters reflect a great deal of education and humor. In one article, "Nellie's Experiment," Cyula describes Nellie's effort to capture a swarm by holding a tree limb. Just when the bees had alighted upon the limb Nellie was holding, the local minister and his wife decided to visit. Eventually, Cyula coaxed the bee swarm to descend into a sheet, but not before the limb broke and Nellie was covered in a shower of bees.[82]

Cyula often commented about the benefits and frustrations of beekeeping for women: "The pursuit is as free to women as to men. There is no prejudice to encounter; no loss of social standing as may be the case in some other employments. The lady beekeeper may expect some manifestations of mild surprise but no disapproval, and in time, if she be moderately successful, she may be greatly surprised that her neighbors exaggerate her modest gains." Although she acknowledges the labor is more intensive for women, Linswick counsels that women should adapt the work to their capabilities.[83]

In the very next issue, Mrs. M. A. Bills heartily seconded Linswick's proposal and cautioned that she needed three years before she showed a modest profit. In the same issue, Mrs. L. B. Baker summarized the limited career choices presented to women: either they had to be teachers, seamstresses, or musicians or they were labeled "masculine" and "strong-minded." She went on to elaborate the importance of women being able to subsist:

> The question of woman's rights is no longer prominently before the public, but whatever one's views may be, or may have been upon the subject, its agitation has undoubtedly done good, leading women to consider her own abilities and awakening her to the realization that whatever other rights were denied her, there were fields of remunerative labor open to her, hitherto unrecognized. These, considering the barriers of custom, she has not been slow to occupy, but there are still others given up to the monopoly of men, to which she is well adapted and which in the progress of woman's or human rights must inevitably be shared with her.[84]

Other women wrote to discuss the health benefits of beekeeping. In an article compiled by W. G. Phelps, Mrs. Jennie Culp of Galena, Maryland, conceded that there were disagreeable aspects to beekeeping, such as stings, hot work, and heavy equipment, but countered that it did "afford that which most of our American women so sadly need—exercise in the open air. I have scarcely a doubt that much of the present debility and physical weakness endured by the female sex of this country would pass away with the increased employment in congenial, out-door occupation."[85]

Even the topic of dress was debated in the bee journals. In 1885 Jennie Culp wrote that the bloomer suit would "naturally" suggest itself as the proper costume.[86] In 1878 Mrs. L. B. Baker advocated that women switch to more convenient dresses when working in the apiary, "for a pursuit which necessitates shabby and untidy apparel is one in which a refined woman will never engage." The solution? Not slacks, which would have been considered far too radical for the Michigan beekeeper. Rather, "a dress that can be made short or long at pleasure." Baker reluctantly acknowledges that bloomers could make tasks in the apiary more convenient and goes so far as to recommend that "the pantaloons should be similar to the ones worn in the old 'bloomer costume'—straight and full, and like the sleeves with rubber cord in the hem and fastened over, not above, the shoes."[87]

Baker advocated wearing dark-colored dresses because beeswax would not look like spots of grease if the beekeeper were to have unexpected visitors in the beeyard. Such advice goes against conventional wisdom because dark colors draw the attention of bees. Even though she states that she herself never visited a sister apiarist, she concludes confidently that if the female apiarist were "so dressed, there need be no fear of bees, and we may, without embarrassment, give a cordial welcome to callers in the apiary or the parlor."

Not all women who wrote to bee journals were keepers. Some, it seems, just wanted to express their sadness at losing the beekeepers in their family. The "Letter Box" in *Bee Culture* and "Notes and Queries" in *American Bee Journal* provided a social forum for those who were often marooned by distance or technology. One is tempted to glide over one woman's article titled "Visiting California Apiaries," written in 1887. Yet, the very first line describes an occurrence that took place in 1885. The author, Mrs. B. Stover of Roscoe, Illinois, wrote that her husband had wanted to see the magnificent California apiaries, so they went, "but the climate did not agree with him." Instead, he grew worse as they continued their trip, until finally in May, he died. Mrs. Stover had him buried beneath a live oak in Ojai Valley, "amid the ceaseless hum of the busy bee, whose music delighted his ear."[88]

The letter ends with a goodbye to a circle of friends who had been important to her deceased husband: "It seems hard to dispose of his bees, and I shall miss them, as they have been a part of the family for the past ten years."

LABOR FORCES

Henry Fauls once wrote, "A lady can take care of ten swarms with less labor than is required to take care of an ordinary lot of house plants," but women strongly disagreed.[89] Mrs. Mahala Chaddock of Vermont, Illinois, wrote that "beekeeping is too hard work for women. When I say women, I mean, of course, American women. German and Swede women can keep bees—they can do anything that requires lifting; American women cannot. All American women have a tendency to heart disease, consumption, or kidney disease, and carrying beehives and boxes of honey is not good for any of these diseases."[90]

Not only did they freely address the physical hardships and differences but the bee ladies were not afraid to acknowledge inequities of their own societies. Chaddock's article provides a list of women beekeepers and their labor force. Mrs. Lucinda Harrison employed Irishmen to do her lifting. Mrs. St. Julienne Moore of Louisiana had black women help her; the Linswick sisters hoisted the hives somehow and got along together. For her part, Mrs. Chaddock complained: "I have no Irishmen and no colored women, and the men are always busy. I want the old colonies carried away just when the bees are working most briskly and that is just the time the men are away off in the fields."

Slave narratives collected in the twentieth century also provide insight into nineteenth-century labor conditions. A former slave named Ann Gudgel was interviewed by Mildred Roberts during the 1930s. Gudgel acknowledged that "the onliest time ole Miss eber beat me was when I caused Miss Nancy to get et up wit dee bees. I ole her 'Miss Nancy, de bees am sleep, let's steal de hony.' Soon as she tetched it, day flew all over us, ant it took Mammy bout a day to get the stingers outen our haids. Ole Miss jest naturally beat me up bout dat."[91]

Chaddock had no qualms about admitting when it was not a good year. Having paid $8,000 for their farm, she attempted to add to the family's income by selling strawberries and honey. "He [her husband, who is not mentioned by name in the article] was willing for me to bind rye all day, or busk corn day after day for weeks, but he thought peddling produce was 'small business,'" she wrote. "But we were so scarce of clothes and dishes, and every thing that I begged him to let me try it." When the railroad was being built close to Vermont, Illinois, she did very well, reducing their debt to $3,000. However, by 1884 the railroad crew had moved on to towns farther down the track, and it was so far to town that she could not justify the use of a horse.[92]

The opportunities that were available in cities, factory work in which women could earn their own wages, the yearning for education—these things made many young women want to leave the farms and start new careers. Mrs. N. L. Stowe echoed many sentiments when she wrote:

> Girls of today are restless and ambitious; rightly or wrongly so, as you will, women are becoming more and more independent; you can not help it any more than you can stop the march of reform in all the phases of life; but you may lead and guide them aright. If you are a beekeeper you can get your daughters interested in your work, and have them help you, and be very sure to give them their pay or their share of the profits, the same as you would any one else. Let them have periodicals, and encourage their little ambitions, and you will, together with them, be lifted up into something better. Surely farmers who are spending

their lives trying to make something grow should see to it that their own lives, and those of their families be not dwarfed.[93]

Stowe's advice to farmers to encourage their daughters to read the bee periodicals emphasizes the point that education of the young can lift the entire family.

Canadian bee women entered the bee conversations via journals published in the United States. Miss Annie Mann started writing in the 1870s. *British Bee Journal* did an excellent job of recording the large presence of Canadian apiarists in the Colonial Exhibition. The statistics recorded do not distinguish apiarists from visitors, although the article says at least a hundred ladies and gentlemen were present. In fact, so many women were present that the keynote speaker commented on the beautiful sight, which was "a proof that beekeepers were loyal to other queens besides the queen bee."[94]

The bee journals reported that the royal family supported and provided research funds to British Isle and Canadian beekeepers. Their Royal Highnesses Victoria and Albert and Prince Leopold and the Princess of Wales financed many exhibits. Princess Beatrice especially endeared herself to English and Canadian beekeepers by sponsoring many honey shows, although the Prince of Wales was equally prominent in sponsoring the Colonial Exhibition and invited beekeepers from New Zealand and Australia to participate. Inhabitants of Norwich were so grateful for Beatrice's presence that they crafted a beautiful bee pin and sent it to her.

If a girl was "restless and ambitious" as Mrs. Stowe says, the periodicals would have given her plenty to ponder. Many of the editorial conventions promoting masculine identity were being discarded. There is one letter in which the editor makes a pointed tribute to those women who did not sign their names but submitted letters and articles to the journals anyway. An article written by Reverend W. F. Clarke, "What Constitutes a Good Bee-Periodical?," discusses the use of the editorial "we":

> Its use is established by the concurrent practice of the best journals in the world. Labouchere and one or two others who have tried to use

"I" are regarded as cranky. Usually there is more than one editorially connected with a paper. Mr. Hutchinson seems to have given in to Dr. Miller's urgency about dropping the "we." Yet more than once he has disclosed the fact that his wife is his best helper, and right hand man. He reads everything to her that is of special interest, and gets the benefit of her womanly justice and instinct. In justice to your noble wife, friend H., do reconsider that rash promise of yours, and continue to say "we."[95]

LIZZIE COTTON: "STILL HUNTING FOR GREENHORNS"

In 1880 Lizzie Cotton wrote *Beekeeping for Profit,* with mixed results and mixed messages. According to historian Karl Showler, there were four printings of this book, although print runs during the nineteenth century were small. She had constructed a hive that ensured regular feeding for bees throughout the year, building the hives up so they could survive winters. She also guaranteed that swarming would not happen, which suggests that she may have been selecting queens for this trait.

Quite a few people experienced hives that did swarm, however, and they were not happy with Cotton. Her name cropped up regularly in the Humbugs and Swindlers column in *Bee Culture.* People complained that their bees *had* swarmed, and they had not made a profit from their bees. Cotton also had a tendency to advertise a sale on her hives *after* the sale date had already expired. "That's just our luck," A. I. Root fumed. "Lizzie Cotton has been offering her $15.00 Controllable hive for seven dollars, but the magnificent offer was only to remain open until Feb. 24th, and we didn't get the circular until March. . . . This hive is not patented; oh no! but if $7.00 gives a profit we Yankees would like to know."[96] In *American Bee Journal,* Thomas Newman published a letter from C. W. McKown, who warned that Cotton was "still hunting for greenhorns who are too penurious to take a reliable bee periodical."[97]

Nonetheless, Cotton stayed in business until the 1890s. In a pamphlet dated West Gorham, Maine, August 5, 1880, she writes:

There is in my opinion no pursuit which offers greater inducements than bee keeping, especially to women. There are very many who are confined indoors nearly the whole time, excluded from the air and sunshine, to the great injury of their health; and after this sacrifice they barely succeed in obtaining a livelihood. To such, bee keeping offers great inducements, such as improved health and a handsome recompense for all labor performed. . . . A lady bought a swarm of Italian bees of me in 1874, and she writes me that from that one she increased her stock to over twenty swarms the third season; besides she got over one hundred pounds of nice honey from the swarm I sent her the first season.[98]

A key point to being a successful beekeeper, Cotton wrote, was managing the drone population. She advocated cutting out drone brood because she felt that they ate too much honey: "I wish to impress strongly on the minds of all who adopt my plan of bee management, the great importance of cutting out drone cells, except a few in every hive. Don't leave more than fifty, it takes a great deal of honey to rear a large brood of drones, and still more to support them in idleness, two or three months."[99]

Just in case there are greenhorns, or beginner beekeepers, reading *this* book, I want to clarify that drones *are* needed in the colonies. There is strong evidence that worker bees will become demoralized without the appropriate number of drones in the hive. Furthermore, in these days of integrated pest management, a nonchemical way to control varroa mites is to remove a frame of drone brood for a few days, put the frame in a freezer so the mites will freeze, and then replace it in the hive. Drone control, if done properly, is a responsible way to manage mite counts. If one were to follow Cotton's advice about drones, the individual hive's health and regional bee diversity would be severely compromised.

And contrary to Cotton's book and her primary sales pitch, there is no hive that controls swarming. This is the worst message that Cotton could have sent to new beekeepers. Swarming is as natural to honey bees as dancing is. There are methods of controlling swarming, but beekeepers have to be knowledgeable about swarming signs in the hive. Even then, if a hive takes a notion to swarm, there is very little a beekeeper can do. Similarly, even when hives should swarm because of overcrowding, sometimes the queen will stay and the workers will continue to produce honey. If beekeepers want to prevent swarming, larvae from those colonies should be used for queen production. Researchers think that the swarming instinct is genetic and that a beekeeper can select for such behavior.

Bees either make honey or they make queens. When they make queens, they are implementing the process of swarming. Swarming can be controlled by creating more room for the mother queen to lay eggs. In this way, a beekeeper can keep the workers busy collecting pollen and nectar and the queen busy laying eggs.

One thing that Cotton was correct about, however, was marketing comb honey. In the nineteenth century, when standards had yet to be adopted by the federal government, good marketing was the key to making money, and Lizzie Cotton knew it. Given that adulterated honey was quite common among unethical honey producers, she went overboard in refusing to consider honey extractors as useful equipment: "[Extracted honey] is expressed in one word: counterfeit. Extracted honey can be easily counterfeited; comb honey cannot."[100]

Although Cotton would be proved wrong about extractors (World War I would make extracted honey a necessity), she had a point: comb honey cannot be manufactured except as the bees construct. There is no possibility for adulteration. Even among contemporary beekeepers, comb honey is an authentic product with no need for certification. But the market for comb honey would end quickly with World War I. According to E. F. Phillips, war rationing (of sugar and the petroleum to run sugar factories) meant that "extracted honey production increased 400 percent."[101]

However, Lizzie Cotton's reputation within the bee industry would suffer because of her claims well into the twentieth century. In part, she never acknowledged that bees swarmed. Another factor seems to have been her prices. One writer, Hiram Adams, wrote to

Bee Culture to say that he had received a good-quality queen from Cotton. Root responded curtly: "With the very large prices that Mrs. Cotton charges for whatever she advertises, she certainly ought to give good measure and good quality, and we are very glad if she is beginning to do so."[102]

Within one century, North American women beekeepers profited from technological advances, especially the movable-frame hive, a new legal infrastructure emphasizing property rights and higher education, and print media, which provided an education that traditional institutions could not. Women beekeepers also took advantage of other technological advances such as the extractor, the smoker, and the foundation maker. They participated in associations and trade journals and marketed their own queens, products, and small-trade journals. Annie Savery used the 1871 Transactions of the North American Beekeepers' Society to promote the bee profession for women and in her speech made a stunning critique of marriage:

> Marriage is too often a thing of mere convenience; when it is so, both are apt to become indifferent, selfish and unfaithful. It should not be so. It should be regarded as a sacred bond—a perfect union of kindred souls—and of all that is divine within us. It should not only be made for time, but should run on through all eternity. This cannot be until a woman can appreciate and claim her God given rights; until she learns to use her powers and faculties—in a word, until she becomes self-reliant. Nothing will contribute so much to make her all this, and to develop her into such woman as every sensible man must admire, as engaging in an employment which will make her his equal.[103]

The record of her presentation does not indicate whether the audience applauded or sat stunned, but it does indicate that at least ten women attended the event.

The rises and precipitous falls of nineteenth-century women beekeepers make for a fascinating tale when one considers that women were disenfranchised in many ways for half the century. By 1851 great strides had been made in three major areas: property ownership, higher education, and technological advances in beekeeping and print media. Even though US federal law would not extend the right to vote to women until 1921, the Married Women's Property Act of New York, 1848 (amended 1860), passed only three years before Langstroth's movable-frame hive was designed. Many states used it as a model for their own laws. For many women who had been unable to make wills, conduct transactions, or rent or sell land as needed, this state law was as much of a high-water mark as the movable-frame hive was to the beekeeping community. The movable-frame hive offered a viable way to improve financial stability, and with both hive and law in place, the United States became a more equitable place for women.

North America

The Great Experiment, Part 2

Women Beekeepers in Industrial Agriculture

I think woman is by birth and training a natural gambler.
—Anna Botsford Comstock, "Beekeeping for Women," 1908

In his book *The Fruits of Natural Advantage: Making the Industrial Countryside in California,* scholar Steven Stoll suggests that five factors merged to create a highly industrialized agricultural landscape in North America at the turn of the twentieth century: unique land conditions, university research and extension, innovative farmers and orchard growers, an independent yet inextricable relationship between farmers and the federal government, and a solid hierarchy between owners and laborers that generally divided along class and racial lines.[1] All of these factors played into a general agricultural trend in the United States to specialize in monocultural crops. Yet, invisible in the industrial agriculture framework were the honey bee and the many women who made and continue to make invaluable contributions to industrial agriculture.

Women beekeepers benefited from well-timed opportunities offered by education, financial stability, smaller families, a well-connected state-federal extension system, and an enhanced legal framework. Canadian women were granted the right to vote in 1918, and US women, in 1921. In these two countries, women beekeepers established their own careers, traveled internationally, earned university degrees, and redefined familial patterns. Mexico, which did not ratify the women's suffrage movement until 1953, did not offer as much economic diversity for women beekeepers until recently, but it too has offered opportunities to women beekeepers.

Twentieth-century women beekeepers took the quickly shifting ideals of femininity in stride. World War I offered women opportunities to become beekeepers because there was a federally supported market for honey and wax. A downturn followed in the 1920s, but in the 1930s and 1940s the United States and Canada fundamentally revised their feminine paradigm to make room for another wartime women's workforce. However, a federal infrastructure supporting beekeepers did not materialize. Instead, cultural and demographic shifts from agrarian to urban landscapes meant a major decrease in beekeeping opportunities for women. Furthermore, as the postwar economy reverted to urban-based locations and large-scale agriculture, women were encouraged to become domestic, facilitating a return to the nineteenth-century true woman. The nuclear family model in the 1950s in the United States emphasized smaller families, a home in the suburbs, and a divorce from agrarian traditions.

From the 1960s on, North American women beekeepers defy easy categorization. They are beekeepers, extension agents, artists, queen breeders, and pollination specialists. This chapter is arranged geographically until the 1960s, at which point it is organized categorically according to Stoll's five factors—federal laboratories, university programs, unique land conditions, educated beekeepers and orchardists, and an exploitable labor force—to show how contemporary women have buttressed the industrial countryside with beekeeping. The chapter concludes with other initiatives that also are successful in North America. The open-ended structure is meant to reflect how women beekeepers can flourish when creativity, flexibility, education, and legal infrastructure are available to them.

Industrial Agriculture
THE EAST

Anna Botsford Comstock: "Mother of the Nature Study Movement."
Anna Botsford Comstock could not be a better example of an accomplished beekeeper, extension agent, artist, and writer. Born in 1854, Comstock was best known for breaking new ground in environmental education, a relatively new topic in academe. Anna decided to attend Cornell University in 1874. She surmised that "Cornell must be a good place for a girl to get an education; it has all the advantages of a university and a convent combined."[2]

But a convent Cornell was not. Botsford met her husband-to-be, John Henry Comstock, when she took his basic entomology class in 1874, and they remained friends through a series of engagements to other people. They married in 1878. Only two years later, John Comstock initiated the first entomology program in the United States at Cornell, and the marriage would define both John and Anna as leaders of a new scientific way of teaching about not only honey bees but other insects as well.

The two formed a powerful intellectual team. In 1887 John Comstock began teaching about bees by issuing a hive to every student; the students were expected to tend to the hives using the lessons of class and practical observation. To provide a backdrop for the Comstocks' new programs, it may be helpful to review a little-known law created by Congress in 1887, the Hatch Experiment Station Act. States were required to provide funding to land-grant universities and colleges to fund agricultural research. Many women such as Anna Botsford Comstock learned beekeeping at land-grant colleges or ended up facilitating extension as their career of choice.

Shortly after the Hatch Act passed, the country experienced a devastating agricultural depression in 1890. In its midst, Cornell University asked the Comstocks to develop a farm program for young people to ensure that they would not get discouraged and leave farm-

Anna Botsford Comstock, New York, United States. It was said of Comstock, she "may not be much of a Christian but she is great as an association." Courtesy of *Bee Culture Magazine*.

ing. Anna implemented an environmental curriculum that would be the model for many schools and universities. In the meantime, John went to Washington, DC, to work on a project dealing with scale bugs devastating Florida citrus crops. The Comstocks provided intellectual leadership during the initial stages of the formation of the industrial countryside in which farmers and orchardists would become monocultural growers, moving away from diversified farms. They split their time between Cornell, the University of California–Davis, and Stanford University.

For her efforts and success in shaping the environmental curriculum, Anna Comstock was promoted to assistant professor of nature study in the Cornell University Extension Division in 1898, but the title—though not the salary—was promptly rescinded after male colleagues complained. Although she certainly equaled if not surpassed her colleagues in terms of longevity of service, she would not receive a professorship until 1913.[3]

Because they always needed money, Anna provided engravings for the Texas Agricultural Experiment Station, one of the foremost land-grant institutions specializing in cattle and onion research. Because the inclusion of photographs would have made books prohibitively expensive for students, many textbooks contained illustrations, and Comstock's were some of the finest. She was the third woman admitted to the American Society of Wood Engravers. Her collaborative work with her husband, *Manual for the Study of Insects,* sold fifty thousand copies in 1926. *Insect Life* (1901) was used by the nature study movement, a discipline popular in the early twentieth century that emphasized exploration and experiments. Comstock's *Handbook of Nature Study* (1911) later came to be called the "Nature Bible." When it came to teaching sensitive topics such as one species eating another, she would focus on the "hungry creature rather than the one that made the meal."[4]

The Comstocks formed their own press to ensure that high-quality scientific work would be available. In one of their editions together, John Comstock gave credit to his wife's artistic abilities. Although John referred to Anna as the junior author, he gave her high praise by recognizing her talent in his preface: "As the skill which she has attained in this art has been acquired during the progress of the work on this book, some of the earlier-made illustrations do not fairly represent her present standing as an engraver. . . . The generous appreciation which the best engravers have shown towards the greater part of the work leads us to hope that it will be welcomed as an important addition to entomological illustrations."[5] Upon the publication of her third book, she was promoted to assistant professor. In 1913 she was finally upgraded to full professor, a rank that her husband had achieved in 1901 with her assistance in publications.

Comstock was a good writer. In an article written for *The ABC and XYZ of Bee Culture,* she addresses the practical reasons why a woman should keep bees: "There is something about the daily routine of housekeeping that *wears the mind and body full of ruts,* even among those who love to do housework better than anybody." She continues in the same style: "If some means could be devised by which housework could be performed with inspiration, zeal, and enthusiasm, the servant problem would solve itself. But this ideal way of doing housework can be carried on only when the spirit is freed from the sense of eternal drudgery.

"I am not a wizard to bring about this change; but I know one step toward it, and this is the establishment of some permanent interest for woman that will pull her out of the ruts and give her mind and body a complete rest."[6]

Comstock's warmth and charm as a teacher drew her to all corners of the earth, including Turkey and Egypt, and to the different pockets in the United States. On a trip to Grand Rapids, Michigan, Comstock was a speaker on the same program as Booker T. Washington. She and her husband subscribed to Washington's Tuskegee Institute, and she traveled to the South to teach nature study to black teachers.

The Comstocks donated their publishing house to Cornell upon their retirement. In 1923 Anna was named one of the twelve greatest women in America by the League of Women Voters. On August 24, 1930, she died, having been awarded a Doctor of Humane Letters by Hobart College earlier in the summer. In 1988 she was named to the National Wildlife Conservation's Hall of Fame. Although she never had children, Comstock finally was acknowledged as the "Mother of the Nature Study Movement."

Tuskegee Beekeeping Ladies with George Ruffin Bridgeforth, Alabama, United States. No date is given, but Bridgeforth was an agriculture instructor from 1902 to 1915. Courtesy of the Tuskegee University Archives.

THE SOUTH

Margaret James Murray Washington. For all her public service, Margaret James Murray Washington, the lady who began the Tuskegee Beekeeping Ladies, remains a mystery. A graduate of Fisk University in 1889, Margaret James Murray married Booker T. Washington in 1892. As "lady principal" and eventually dean of women, she worked with her husband to make Normal and Industrial Institute at Tuskegee, Alabama, a respected school emphasizing practical vocational skills. In her role, Washington taught canning, sewing, hygiene, child care, and literacy. If black women had domestic skills, the philosophy went, they could always find employment. Her Beekeeping Ladies received national attention for their work in the hives. The lasting legacy of this class is a photograph in which instructor G. R. Bridgeforth is surrounded by young women peering over a frame.[7]

A nationally recognized leader in women's rights, Washington also believed that "Tuskegee Spirit" could help ease the South toward reconciliation with interracial conflict. Her approach was straightforward. In an address to the Old Bethel AME Church in Charleston, South Carolina, on September 12, 1898, in which her audience was composed primarily of middle-class black women, she plainly called for a new social morality. She wanted women to take responsibility for unplanned pregnancies, poor diet, and poor hygiene.[8]

Everything in her life pointed to creating a new environment for black women, and that included preparing women to be beekeepers. "Plain talk will not hurt us," she said in her Charleston address. Unfortunately, whereas Anna Botsford Comstock had abundant writings about honey bees, in large part because she and her husband started their own press, Washington was not well published and her efforts to establish solid beekeeping programs in a very difficult culture have been forgotten.

THE HEARTLAND

Women Honey Producers: The Royal Road to Wealth. Many women turned to beekeeping during World War I when men were fighting in Europe. Government programs encouraged honey and wax production, and many teachers supplemented their incomes by becoming beekeepers, or in some cases made permanent career transitions. "Because of the sugar shortage, thousands began beekeeping," explained extension specialist E. F. Phillips, writing in the 1940s. "During the first war, there were abundant means of spreading information about sound beekeeping practices through bulletins, extension workers, and short courses. . . . There were no price controls, which led to inflated prices for their crops, which led to speculations and the incurring of heavy debts."[9]

In 1916 Belle McConnell of Des Moines, Iowa, was asked why she went into bees. She replied: "I think it is good for a person to be interested in some line other than that by which he earns his living."[10] This remains good advice today.

A teacher by training, McConnell decided to begin beekeeping during a summer vacation. Upon getting her hives, she writes: "I spent a great deal of my spare time walking in the yard and inspecting the outside of the hives. I had not that insatiable desire which so many beginners feel to look into them. I was satisfied to let my newly acquired pets proceed with their housekeeping unmolested."

The metaphor of domesticity continued in her report: "Old beekeepers say it was a freak year and bees did unusual things. It seemed to me that by ten o'clock they were issuing from every entrance, and the air was filled with them. I produced hives of the newest and most approved styles, and tremblingly offered them to these adventurers; but one and all they scorned them. Perhaps they found the plumbing bad, or the sleeping porches were not to their liking. At any rate they would very soon sally forth and go back into the old home or into some crowded neighbor's house." Her second summer was a little more successful: "I never had a disgruntled tenant the second summer so far as I knew."

The best benefit McConnell enjoyed was the "vision of a broader life and learn[ing] to lay aside the cares which fret and the trials which harass, and bring the gray hairs." Belle's picture in the report of the Convention of the Iowa State Beekeepers' Association shows very few gray hairs.

In nearby Illinois, Emma Wilson, sister-in-law to well-known beekeeper C. C. Miller, once had "no thought of anything else" but being a schoolteacher. However, when a doctor recommended a sabbatical, she complied. *American Bee Journal* editor Frank Pellett describes Wilson as "too much of a student to spend her spare time in idleness." However, in assisting Miller with his bees, "[Wilson] forgot her resolution to return to the schoolroom at the earliest moment possible."[11]

Instead, Emma and her brother-in-law prepared for comb honey production, having a banner year in 1913 that broke honey records. She also wrote a column called "Our Beekeeping Sisters" for *American Bee Journal*, started in the early part of the century. Editor George York explained its purpose: "It is natural that a woman can offer the suggestions along this line which her own experience finds worthy of repeating better than can possibly come from a man."[12] Wilson's column covered a wide range of topics, from recipes to field tips to organizing beekeepers. The column was discontinued, however, as women's dress changed to include trousers, and smaller, more urban families meant that women were not cooking as much. When Emma Wilson died on April 1, 1933, her death so closely followed her sister's that many journals listed their obituaries together.[13]

In Wisconsin, Mathilda Candler started with a couple of hives in 1917. After a failed attempt to follow her dream of being an artist, Candler returned to her hometown, Cassville. Soon, she had forty-nine colonies. In one year she harvested a crop that sold for $600. According to Pellet, "[This crop] started an epidemic of beekeeping in the neighborhood. Everybody went crazy about it and decided that it offered a royal road to wealth. One large land holder shipped a carload of bees and located them near her apiary. As a result, nobody got much honey for a time, and it was not long until the carload of bees and equipment had dwindled to junk. Few of the fortune hunters stuck, and soon Miss Candler had the field to herself as before."[14] With the wartime economy driving the demand for beeswax and honey and the cosmetic industry needing beeswax to meet the postwar demand for lipstick, women beekeepers stood to make at least as much money as schoolteachers, and with presumably much less stress.

When World War I began, themes of democracy and patriotism started to appear in women's contributions to bee journals. Trade unions began winning some of their battles for labor laws with strikes and negotiations. Although beekeepers tried several times to create a union, most notably in Chicago and California, they never succeeded. In the letters written by women, one detects impatience with labor parties wanting social justice, including women protesting for the right to vote.

Writing in 1919, Mrs. Armstrong Allen noted that beekeeping "has brought no problems to add to the increasing complexities of modern industrialism. Labor has no long score, no aching grudge, to settle with us. . . . And we are producers of a food stuff of real value. Moreover, no store or bank, not even the boasted sunlight factories could be made as hygienic as our apiaries. The worth-while beekeeper keeps them so."[15]

The words of an Iowa beekeeper, Mrs. Clara T. Noel of Oskaloosa, fairly ring with patriotism: "When supplies are scant, the bees go on scant rations and no one murmurs; their stores are so carefully guarded, supplies issued in such a just manner that the colony often lives through a severe famine. But when the harvest is ripe and white clover in bloom, how the bees do work! They let no opportunity pass to bring in supplies even from the humblest source. . . . There are no strikes for shorter hours. There is not a slacker in the hive, no one asks exemption from the duty at hand." Noel ends with a note of optimism common to World War I: "The future need of the colony is at stake, the life of their home depends upon each one doing her bit, and all cheerfully rally to the call."[16]

THE WEST

California Queen Crews. Just as industrialism changed labor relations, so too did industrial agriculture change the bee industry. In writing about the successful queen producer M. H. Mendleson of Ventura, California, Frank Pellet focused on several key points that made Mendleson successful: he pollinated lima and bean fields, he requeened his hives, he raised his own queens, and he maintained an all-female crew for the fourteen-hundred-to-two-thousand-colony queen-rearing yard.

Neither Mendleson's queen grafter nor any of the female crew is mentioned by name, but his opinion was that "women are well adapted to queen-rearing as they are careful and painstaking, and there is no heavy lifting connected with this particular work."[17]

Radio kept people connected, providing quick commentary on the social movements of the day. *Uncle Josh,* as recorded by Cal Stewart in 1919, was a popular post–World War I radio program. In the classic tall-tale tradition, his show about suffragettes begins with an easygoing fellow named Jim, who was known for driving a swarm of

Mendleson queen crew, California, United States. Mendleson was one of the first to hire an all-female crew to graft larvae. This crew is wearing their "farmette" suits. Courtesy of *American Bee Journal*.

bees from Maine to California and back again without losing a bee. His trick: "You had to get some old bee that was halter-broke and would stand without hitchin'," Uncle Josh says slowly, "and put a bell on her." *(Pause.)* "Where she went, the other bees would follow."

Finally, intrigued that his friend could control bees the way he would cattle, Uncle Josh took the bait and ordered himself some bees. He received six hives of bees and a book about the bees, especially the queen. "I wrote them for a king bee," Josh says rather heatedly. But apparently, the bee supplier stood his ground and told him that he had suffragette bees, "which wouldn't have any king bee around."

Inevitably, the suffragette bees swarmed, and in spite of Josh's fancy equipment, "they just sot up in the top of a tree and wouldn't come down." That convinced Josh that these weren't suffragettes, these were International Workers of the World bees who had decided to go on strike.[18]

The Kerrs. Uncle Josh was mild compared to other critics of suffragettes. There was no doubt that women beekeepers exercised an economic presence inside the home, if not a political presence outside of it. In California, Lela Kerr was married to Russel Kerr throughout the 1920s. As with many women beekeepers, Lela maintained one foot in the classroom and one in the beeyard, helping her husband in their migratory beekeeping business. The Kerr family had been renowned in the San Diego region for their bee ranch, two hundred acres and six hundred hives, which they had bought from a homesteader named Ira King for $400. James Fenimore Kerr, Lela's father-in-law, had two mottos: a fool and his money are soon parted, and if the fool is his son, he (James) might as well take the money as anyone else.[19]

Russel wrote an autobiography that detailed his father's difficult nature. He also carefully recorded Lela's efforts to keep them out of debt, as well as his daughter Elizabeth's efforts to help with the bees: "My daughter was a very good helper in the bee yard, but the bees had a habit of stinging her on the seat of the pants when she stooped over—that straightened her up real quick."[20] Both Lela and Elizabeth were good in the beeyard and the extraction room. As if having a difficult patriarch were not trouble enough, the Kerrs had to deal with arbitrary weather patterns. Russel Kerr's bee journals record wild swings during the drought years.

But the weather was just one obstacle. The Kerrs specifically and beekeepers generally tolerated exorbitantly high railway freights, tariffs, and honey exchanges. Before 1924, many European countries imported California honey, including Germany, England, Belgium, Holland, Switzerland, and France. But the market dried up after World War I.

E. F. Phillips summed up the difficulties that the Kerrs faced: "While the general slump came in 1929, beekeeping was caught in

Russel Kerr's Bee Journal

One look at the dates in Kerr's journal shows that adjacent years had opposite effects: The dry winter and low honey crop years were 1883, 1893, 1903, 1914, 1924, 1934, 1944. The wet winters and bumper crops of honey were in 1884, 1894, 1904, 1915, 1925, 1935, and 1945. For every drought year in which bees starved, a wet year followed in which the Kerrs could produce honey crops.

the minor depression of 1921. Producers had overbought and overexpanded; packers had bought honey at high prices and saw it decline on their hands; manufacturers had extended unwarranted credit and had overbought lumber and other materials. When things went bad, cooperation among beekeepers virtually disappeared and there was not much development of leadership in the industry."[21] The financial troubles forced Lela to return to the classroom. She taught until 1924, when they were able to crawl their way out of debt.

THE NORTHWEST

Lillian Hill. Some women beekeepers managed to make their own apiaries a classroom. Mrs. Lillian Hill in Washington State had the good fortune to make sixty-five hundred pounds of honey in 1914 from twenty-six swarms of honey bees working fireweed flowers. The next year, most of her bees died as a result of European foulbrood. Yet, she found that bees were a good way to reach young people, whom she often employed or offered shelter. "She discovers her boys and girls everywhere in reform schools, orphan asylums, on the neighboring farms and in the city," wrote Harriet Geithmann in an article on Hill.

In 1924 Hill invested in thirty more hives and loaned them to a girl to help her earn money to pay tuition to attend school. The girl lived next door to an abandoned schoolhouse on an acre of ground. Eventually, Hill rented the schoolhouse and used it to board young people who wanted to learn beekeeping. One young man rode a bicycle from Colorado to Washington to work with her. By the time he had split a fifty-fifty commission with Hill, he cycled away four months later with $500 in his pocket.

The article makes no mention of disasters other than the foulbrood. "I never camp on either the trail of my successes or my failures," explains Hill. "I go right on."[22]

THE NATION DURING THE GREAT DEPRESSION

In addition to the massive environmental damages affecting the nation, the 1930s brought two federal changes pertaining to women beekeepers: the massive social relief programs for underprivileged and unemployed families, and the 1938 Food, Drug and Cosmetic Act. These federal initiatives, along with loans from the Farm Security Administration, supported honey and beeswax products; but in the lean economic times, serious flaws in the social safety networks were exposed and not addressed until after World War II and well into the 1960s.

The National Honey Institute was especially important in trying to generate honey sales by educating the masses. The institute worked with civic associations to offer cooking demonstrations using honey. The Iowa 4-H Club program was exceptionally successful at marketing honey to young women ranging in age from twelve to twenty-one. In each club meeting, Iowa beekeeper Lulu Tregoning discussed honey production in the young women's own communities, sources of honey, and honey recipes. More than eighteen hundred girls and two hundred leaders participated in these classes. Tregoning continued teaching these classes for at least four years, although once World War II began, her columns ended.

Inevitably, faced with declining honey sales, beekeeping families did what they have always done: the women returned to teaching until the economy turned around. Woodrow Miller's sister Florence would help out with the accounts during the summer and would teach in Provo. Quoted in Rita Skousen Miller's family history, Woodrow Miller sums up the situation succinctly: "Beekeepers who could teach school went back to schoolteaching; they tended their colonies with a minimum of work, just keeping them from starving and trying to keep them alive with little attention."[23]

Women could work with beeswax. Unlike honey, which dropped precipitously in price, beeswax increased in value during the Great Depression. According to Woodrow Miller, "Most the price of honey had gone down to as low as 3 to 4 cents a pound. The cost of producing honey at that time was approximately 5 cents a pound. The price of the byproduct, beeswax, had gone to 10 cents a pound in 1932."

The steadiest markets were candles and cosmetics. Fortunately, these markets hired women factory workers, even though "there is no sane connection between morals and cosmetics," according to Nell Vinick. Beauty products, Vinick contended, were "merely symbols of the social revolution that has gone on, the spiritual and mental forces that women have used to break away from conventions to forward the cause of freedom."[24]

Even as the Great Depression meant declining discretionary spending, cosmetic companies marketed lipstick and skin creams as "food" for the skin, and the sales of these items increased steadily through the 1920s and 1930s. Since jobs were scarce, young women took beauty classes, thinking that cosmetics might give them an edge in the job market. According to Kathy Peiss, "Most families spent only small amounts on personal care items, about 2 percent of their income, but this added up to a total outlay of $750 million nationally in 1931." Those sales also increased through the 1940s, even as women were rationing gas, sugar, and other commodities.[25]

The advertisements for cosmetics became so misleading that consumer advocates successfully lobbied for federal legislation, resulting in the 1938 Food, Drug and Cosmetic Act. Before this legislation, one cosmetic, Koremlu, contained rat poison. Another product, Lashlure, an eyelash beautifier containing aniline dye, blinded and disfigured a young society woman from Dayton, Ohio. While many women were happy to exercise their right for the legislation requiring truth in advertising, they would not give up their "hope in a jar." Cosmetics offered women a "language," as Peiss describes it, a rhetoric of self-expression that was socially accepted.[26]

Being reined in marginally, the cosmetic companies saw World War II as a marketing gambit that had to be carefully trod in order for women and the government to agree to its beeswax needs. Patriotism came in "themed" cases, specifically lipstick. One company marketed lipstick as a type of bullet. Lipstick became a conflated symbol of freedom: women's personal freedom, national freedom, freedom from rationing. When the War Production Board met to decide the necessity of rationing cosmetics, the one item that went completely uncontested against rationing was lipstick.[27]

So beekeepers sold beeswax throughout World War II, although they did not make much money compared to the beekeepers during World War I. E. F. Phillips wrote in a state apiarist report filed in 1942, "There has been an increase in the use of honey by industry, especially cosmetics and lately by the products of defense." The War Procurement Board imposed ceiling prices on honey and beeswax. Although these price controls were meant to curb the wild down-swings that happened in the 1920s, they also dampened enthusiasm for entrepreneurial women. Supplies were difficult to attain with shortages of nails, lumber, and paint.[28]

More important, the federally supported extension services were not in place for beginners. "The short courses that were so effective in the first war have not been undertaken at this time," Phillips wrote, "although the need for them is greater now than before." Many states restricted these courses because of gasoline rations. They also restricted extension services, preferring to use radio as a means to disseminate information. During National Honey Week in 1940, the Iowa State College radio station promoted honey recipes in the "Homemaker Half Hour," led by Elizabeth Kraatz, but radio proved no substitute for honey-cooking shows and samples of baked goods. To sum up his critique of post–World War II beekeeping, "The out-standing lack of beekeeping at the moment," Phillips admonished no one in particular, "is that of leadership."[29]

Whereas in earlier Iowa reports women were active correspon-dents, association presidents, and prominent beekeepers, no women writers were included in the Iowa apiarist reports from 1939 to 1945.

The entertainment industry did its best to hypnotize its audi-ences grappling with the Great Depression and World War II with idealized versions of femininity. If movies could be social narcotics, then bee charmers were one of the few agrarian images that could soothe worried minds. For instance, in *Camille* (1936), Greta Garbo, playing the title role, leads a group on a tanging expedition. In one of the first movies in which actors and actresses spoke, Garbo soothed audiences, acting as a mediator between people and an industrial landscape.

POSTWAR LANDSCAPE

Roberta Glatz. The reality of being a woman beekeeper was another matter altogether. New York honey producer Roberta Glatz started helping her father with his hives during the World War II years. Her mother was an active leader in the Girl Scouts. Glatz's initial interest had been to be a biologist, but the labs interfered with field hockey practice. She had always been good in languages, and in those days, women were either teachers or nurses or retail clerks. She became a foreign-language teacher. In 1949 she married, and in the course of several years, she had three children.

Roberta's reentry into bees came when she was traveling in Germany on a Fulbright fellowship. She met a German woman living in the Alps who had bees. This event stayed with her. As her life's path unfolded, Roberta needed to support her children and remembered the German beekeeper. Glatz looked around her to see what she had: eighty acres of forest, swamp, and a steep hillside, not marketable for timber or agriculture but perfect for keeping bees.

In 1958 Glatz started with two hives, eventually ending with 125 hives to supplement her teaching income. For someone who had to juggle a family and a teaching career, Glatz found that beekeeping had two assets: she could do it from home and around her schedule. April and May were tough times because of swarming and teaching responsibilities, but everything worked out. Sometimes the dishes didn't get done. However, one daughter developed a love for beekeeping, and she keeps bees in Italy today.[30]

After she retired, Roberta did not want to just quit, so she volunteered to do research for preeminent bee specialist Roger Morse's native bee study. Farmers were having a tough time with the pumpkin crops in 1994, and Morse organized a committee to help. At the end of the program, he asked each member to write a chapter contributing to a book written from the point of view of what the farmer needs to know. Roberta focused on a native ground-nesting squash bee, *Peponapis pruinosa*, which has the following characteristics: it collects pol-

Solitary Bees

Compared to honey bees, which have huge numbers, are easily transported, and are relatively docile, solitary bees, according to Roberta Glatz, are "never noticed . . . unless they are eating something or biting you." Solitary bees have advantages such as being specialists and having less competition. If parasites are problems with social bees, solitary bees have an advantage because fewer are affected. The nests are generally underground, making them harder for predators to find.

len and nectar, mates overnight, and cruises flowers to find a mate.[31] Nothing had been done on the native pollinators because they are wild and thus not attractive to commercial companies. "No one will make money," Roberta said matter-of-factly when I interviewed her.[32]

Roberta explains that research scientists have concentrated on honey bees and pollination. Only a few researchers focused on native pollinators: "They are rare people and never got out of the scientific journals."

Just in the state of Delaware, where I interviewed Glatz, there are fifty-seven species of plants that native pollinators will forage: willow in March and April; milkweed in June and August. Goldenrod is a key floral bridge between mid-season and late-season bloom.

In shifting from commercial beekeeping to research, Glatz transferred all her teaching skills to studying solitary bees: how to research, how to judge validity, and how to work with commercial beekeepers, who tend to be male. Her experience as the first female president of the New York Beekeepers Association, a group that has met for 140 years, was also helpful. The men kept trying to throw her out, but she wouldn't let them: she became president. "That was my revenge!" she laughs.

Having served on more than enough committees myself, I replied casually, "Perhaps letting you be president was *their* revenge."

"Either way," she quipped, "we both ended up happy."

Industrial Apiculture

The 1960s solidified several factors affecting women's employment, education, and international experience. In 1961 President John F. Kennedy implemented the first affirmative action laws, addressing equity in the work environment with Executive Order 10925. It was written "to ensure that applicants are employed, and that employees are treated during employment, without regard to their race, creed, color, or national origin."[33] Four years later, President Lyndon B. Johnson would expand that executive order to include gender. The merits of affirmative action laws continue to be debated, but in terms of changing the environment in which women could apply for state and federal positions, the laws have been effective. Although discrimination would continue to be a factor, federal bee labs and state universities became more accessible for women beekeepers after 1961.

Many women beekeepers had options to do international beekeeping in the Peace Corps, which gave them invaluable experience that transferred to the United States when they returned. Other women beekeepers joined entomology programs. Quite a few never left North America, finding that with the federal law, there were economic opportunities in the United States. Even now, women beekeepers serving in universities and federal bee labs provide invaluable leadership and collective experience as North America transitions from a pre-varroa world to a post-varroa world to the colony collapse disorder world.

FEDERAL BEE LABORATORIES

Since the 1960s, US federal agencies have acknowledged affirmative action laws, so women and minority applicants have had more equitable consideration when applying for research positions. In addition to complying with federal guidelines, the US Department of Agriculture (USDA) focuses attention on pollinator health. Currently, there are four overall objectives of the research: healthier colonies, mite management, African honey bee management, and pollination.

Of the USDA's 250 labs, there are only four bee labs—Beltsville, Maryland; Baton Rouge, Louisiana; Weslaco, Texas; and Tucson, Arizona—reduced from the eight the government maintained until 1995. (There is also a lab in Logan, Utah, devoted to native bee problems, but that is not within the scope of this book.)

Despite the decrease in the number of labs, women bee researchers have shown wide-ranging leadership in the new fields shaping beekeeping industries, such as nutrition, integrated pest management, genetics, and diverse bee species and pathogens. In short, the federal bee laboratories synthesize research for honey bees, canvassing the world in an effort to establish international approaches to serious pathogens, viruses, fungi, nutrition, and analysis. The idea is that with the bee labs working in cooperation, the country can move quickly to address sudden losses. The USDA bee labs are, first and foremost, problem solvers, says Anita Collins.[34] Researchers such as Collins, Diana Sammataro, Lilia de Guzman, and Gloria diGrandi-Hoffman represent a continuity of experience at the four federal laboratories.

Anita Collins. From the 1970s, Collins crisscrossed the United States, working in three USDA labs before she retired. In 1976, the year she obtained her doctorate from Ohio State University, she started her research with the agricultural research station (ARS) in Baton Rouge, which focuses on bee stock. This was a good fit for Collins because her doctorate had been on sterile drones. Opened in 1928, the Honey Bee Breeding, Genetics and Physiology Research Unit maintains genetic lines for resistance and management. In an effort to understand how African honey bees would change North American beekeeping, the Baton Rouge team carried out intensely planned research that lasted throughout Collins's tenure at that station. From 1979 to 1989, Collins collaborated with the Baton Rouge team as they went to Venezuela to study African honey bees. Every year, the team took two trips, one before Christmas and one after. Collins initially hoped that drone saturation would be a factor affecting African honey bee temperament, but she learned that that was not an answer. (For more details of this stage of her professional development, see the South America chapter.)

Collins stayed at the Baton Rouge station for twelve years. During this time, the lab began to focus on issues affecting other honey bees besides the African honey bee. Her colleague John Harbo, now retired, worked to develop the Suppressed Mite Resistance genetic line. Another colleague, Tom Rinderer, initiated the Russian honey bee project (in which Lilia de Guzman assists), selecting for multiple traits such as hardiness and mite resistance and maintaining breeder quality. And the lab continues to monitor tracheal mites because southern colonies tend to be susceptible. The lab wants to provide some oversight for genetic resistance among queen breeders in the South.

In 1988 Collins moved to the lab in Weslaco, Texas, where she was "isolated from everywhere, [and] loved it."[35] The USDA opened the lab to conduct research on citrus and vegetable processing in 1931, but the affiliated Honey Bee Research Unit at Kika de la Garza Subtropical Agricultural Research Center has expanded its studies to include much more. Nestled in the heart of the Rio Grande Valley, the Kika de la Garza lab monitors bee nutrition and the movements and migrations of African honey bees and fire ants. In addition, it uses aerial remote sensing of agricultural problems, such as the migration of African honey bees into urban areas.

The Weslaco lab needed a public relations person who had experience with African honey bees. Because of her time in Venezuela, Collins was a good fit. She served as a liaison with the corresponding Mexican agricultural offices. She also did some research, primarily along the lines of finding deterrents for African honey bees to be used by water-meter readers and agriculture workers. But in terms of transferring her research in Venezuela to Mexico, she found the African honey bees there were much more defensive than the bees in Venezuela. This experience reaffirmed the shift in focus that happened while she was at Baton Rouge: bee breeders needed to keep

colonies as Europeanized as possible by requeening rather than selecting for gentleness among African honey bee colonies.

In 1996 Collins moved to the Beltsville Lab, a great move for her
because she shifted the nature of her research to the preservation
of bee lines. Earlier, in 1992, Congress had mandated that the ARSs
begin to focus on genetic diversity among those species related to agriculture. New technologies to evaluate drone semen had developed
since Collins had worked on her doctorate. "So, you could look at
drones and queens in very different ways," says Collins, with audible
enthusiasm. She spearheaded frozen plasma preservation.

The Honey Bee Genome Project has the potential to solve a number of questions about queen bees, such as semen storage feasibility.
Whereas other scientists are interested in pollination or pheromones,
Collins is focusing on what is in the sperm duct. Queens will mate
with a dozen or more drones on several mating flights. A queen will
hold only 10 percent of the sperm in its ovaducts. Even though the
sperm are mixing, they still have to swim to the queen's spermatheca,
a process Collins describes as "rushing up the duct to get 'the best
seat in the house.'" But Collins finds that despite the sperm's mixing,
she still tends to get clumps of sperm, so it is not entirely randomly
distributed. "You want that mixing to get genetically diverse workers. . . . It is important to the hive's survival. If workers are too much
the same, it cuts down on the hive's efficiency to meet challenges of
temperatures or stress," Collins explains.[36]

When discussing women's equity issues in the federal research
environment, Collins felt that the federal bee labs genuinely respected women's contributions because there were already women in
established positions in the bee community. Aside from a professor
who refused to serve on her doctoral committee because she was a
woman, Collins had good experiences with her mentors. The highly
regarded researcher Walter Rothenbuhler was her major professor.
"Rothenbuhler didn't have a discrimination gene whatsoever," she
says definitively.

In Collins's experience, the ARSs formalized the evaluation process so that the emphasis was on the scientists and their impact on the community. Being one of two women scientists in Louisiana in the 1970s, Collins served as a contact person for affirmative action cases. The federal government trained her to be proactive and opened her eyes to some practices that would be unacceptable. This training proved especially valuable in Texas, where women sometimes did not share the same status as their male colleagues.

Although she has retired from the federal system, Collins continues her research with various groups. As a member of the native bee survey group coordinated by Sam Droege, United States Geologic Survey, she traps and identifies native species in the Pocono Mountain and Lehigh Gap areas of northeastern Pennsylvania, as well as in Berks County, Pennsylvania. Recently she was appointed an adjunct professor in the Department of Entomology at Pennsylvania State University. "If I had another lifetime, I'd go another thirty years," she says.

Diana Sammataro. On the other side of the country, the Carl Hayden Bee Research Center in Tucson, Arizona, conducts research to optimize the health of honey bee colonies through improved nutrition and control of varroa mites. The goal is to help beekeepers maximize production of honey bee pollinated crops. As a researcher, Diana Sammataro covers various topics, from bee nutrition to floral sources and probiotic work that may lead beekeepers into new ways of treating colonies.

I was first introduced to Sammataro in 1998, when I opened a Christmas gift in Philadelphia. Standing out amid all the other festivities going on around me, *The Beekeeper's Handbook* by Diana Sammataro and Al Avitabile was the first bee book in my library, and I still cherish it after all these years.

One look at *The Beekeeper's Handbook,* at least its first three editions, and one knows immediately that the authors see the world differently. The book uses a landscape format, with the illustrations placed beside the text on the same page. Readers could therefore os-

Diana Sammataro, Arizona, United States. Courtesy of Randy Cary.

cillate between the text and the diagrams that Sammataro drew. As strange as it seems in our graphic-saturated world, before the 1970s, not very many bee book authors included illustrations with their directions. Pictures, typically unlabeled photographs, mainly appeared as insertions in the middle of the text. But the landscape layout of *The Beekeeper's Handbook* opens up a wider area, so a beginner can take the book to the field and use it like a manual. The book lays out flat on top of a hive or the back of a truck bed. The first edition was designed with a lot of white space precisely so that people would make notes in the margins. The authors included several options for tasks rather than presenting a one-size-fits-all beekeeping approach.

The fourth edition of *The Beekeeper's Handbook* has a more conventional 8-1/2-by-11-inch layout. Its basic premise, however, is still the same: there are many ways to do tasks in the beeyard and the beekeeper must be flexible and receptive to the information the honey bees provide. With this simple approach, Sammataro and Avitabile reversed a trend in how-to bee books in which beekeeping was taught as a mechanistic process.

Sammataro's unique vision may have come from her family. One side of her family hailed from Italy, one side from the prairie heartland in the United States. She grew up working beehives with her grandfather George Weber in Arrowsmith, Illinois. Her mom was a landscape architect and her dad an architect. Sammataro also has a degree in landscape architecture, but her goals were different from her mother's. She wanted to specialize in prairie/wetland restoration. However, the entire culture of landscape architecture at that time was focused on managed environments such as golf courses and residential developments.[37]

She returned to the States from a Peace Corps assignment in 1981 (see the Asia chapter), and in 1984 Sammataro happened to have a conversation with esteemed bee researcher Erik Erikson, who was then at the ARS in Madison, Wisconsin. Erikson hired her. At that time, the ARS lab worked with genetic stock of honey bees, but Erikson had just published a paper on electricity and pollination. Sammataro immersed herself in sunflower varieties and pollination. In her position, she used electron microscopes to study the exterior of honey bees and the pollens they attracted via tiny electrical currents, as well as the different shapes of sunflower nectaries. The ARS lab closed that year, and the beautiful bee garden in which she studied is "long gone."

In using photography to study bees and their relationships with sunflowers, Sammataro began to understand the importance of floral nutrition in bee pollen and nectar. She also began to have serious reservations about the commercial sunflower business. "What are we doing to the bees when so much emphasis is placed on sunflower oil?" she asked rhetorically when I spoke with her. It is a question so basic and simple that it lingered in my ears hours after our conversation had ended. "Some of the sunflower varieties don't even resemble sunflowers anymore!" she remarked.[38]

Two years ago, Swedish scholars approached Sammataro to collaborate on studies regarding beneficial microbes that honey bees produce in their crop, their second stomach. Such collaborative interest has led Sammataro into a new field. Metagenomics studies how genes work in relation to the environment. Sammataro's work now is focused on how the bees' diet affects these microbes. The growth of microbes in a colony is largely due to incoming pollen and nectar, which both contain numerous bacteria and yeast. Since so many beekeepers in the United States use antibiotics, and so many farmers and orchardists use pesticides, Sammataro wants to see just how antibiotics and pesticides work against the honey bees' naturally occurring bacteria.

An article Sammataro coauthored notes: "The latest research suggests strong associations between the microbial community in a person and the occurrence of various digestive tract diseases and even obesity. We might be able to make similar associations between microbes associated with different lines of bees and how it translates into factors such as disease tendencies, worker longevity, overwintering success and colony survival."[39] This new direction leads Sammataro to face one of the main challenges for future scientists. Since so much of US research is funded by grants, which are generally short term, scientists have a difficult time doing systematic research. "Think about Marie Curie," Sammataro implores. "She spent a lifetime doing her work on radium and radioactivity."[40]

Sammataro's greatest strength is that she continues to see the world from the "landscape" rather than the "portrait" perspective. Sometimes she can see from the viewpoint of a honey bee, sometimes from that of the flower. Sometimes, she just recognizes when someone else sees a new angle. Rather than complicate that vision, Sammataro simplifies science so the rest of us can wrap our minds around it.

Lilia de Guzman. Lilia de Guzman has been at the USDA bee lab in Baton Rouge since 1989, first as a student, then as a temporary employee, and finally as a permanent employee of the lab. In her current role, she is a team member of the Russian honey bee project, which provides federal oversight to monitor the introduction of Russian honey bee genetics into the United States and the distribution of the genetics via selected queen breeders throughout the states. Russian honey bees have shown a substantial amount of resistance to varroa mites. This program seeks to provide more genetic diversity in the United States and another measure of varroa mite control to beekeepers.

De Guzman, who was born to a poor agricultural family in the Philippines, sees similarities between the Philippines and Louisiana, particularly in how women are socialized: "There's too much poverty in Louisiana. I don't know what may be the main reason or underlying factors to so many uneducated women in the state. I think this lack of interest starts when they are young. A lot of school children do not excel in science and math classes. These classes just scare them. Of course, lack of help or guidance from their parents, who are often uneducated too, does not help."[41]

Speaking specifically about women beekeepers, de Guzman observes that in general, "a lot of women are afraid of getting stung by honey bees. This alone will deter them from becoming a bee scientist." De Guzman also points out that another cultural constraint among southern women is how very family-oriented they tend to be. "This may hinder southern women from pursuing graduate school," she suggests. "School work and research take a lot of time. Hence, it would take a lot of time away from their children. I also suspect that fear of having a divorce may also be an important driving factor, since lack of quality time with their family will eventually cause a strain in their relationship. There are also those who are worried that having a higher degree than the spouse will cause insecurity that will later hurt their relationship. They just don't want to take the risk."

De Guzman knows firsthand the stresses of balancing graduate school, family, and work. "I gave birth to my first child Sheila when I was finishing up my master's degree at Oregon State University in 1989 under the supervision of Dr. Mike Burgett," she relates. "Being a graduate student, I was not entitled to all of the benefits that a regular employee gets, such as maternity leave. I was very scared and worried how we were going to survive. Nevertheless, taking time off to take care of my newborn was not really a problem. But I did not take advantage of my professor's kindness; I went back to work after a month. . . . My husband was just making the minimum wage at that time so we didn't have enough money to pay for daycare."

De Guzman remembers, "My first baby was five months old when we moved to Louisiana for me to work on my doctorate at Louisiana State University under Dr. Tom Rinderer. Being a student and having a toddler were very difficult tasks. I don't know why, but it felt like every midterm exam my daughter was teething." So using the creativity she learned as a child in the Philippines, de Guzman studied "with my daughter on my lap and a book in front of me, reviewing as much as I could while I rocked her to sleep. My grades were not all A's but they were A's and B's."

Never one to stop, de Guzman plunged ahead to her dissertation. "My research on the Yugoslavian honey bees required a lot of traveling to Grand Terre Island and then to Florida about every 10 days because varroa mites were not in Louisiana at that time," she recalls. "Not only did I miss one semester of work but I missed my family terribly while on travel, especially my daughter."

Then, "after almost four years of school work and research in Florida," de Guzman became pregnant again. "This time, there were two of them! Yes, I was screaming and very worried at the hospital when I learned that I was having twins, a boy and a girl! Letting my professor know that I was pregnant was terrifying. I was worried. I did not want to let him down. But he was very supportive. Complications are expected when having multiple births. I was on bed-rest for six weeks but took advantage of it by writing chapters of my dissertation."

Fortunately, de Guzman was not rearing her children by herself: "In order to stay focused with school and research, my husband did all the chores at home while taking odd jobs to support the family." Nevertheless, she says, "with little money and no relatives to help, it was tough raising one child, much less raising three. I remember tearing up so many times while preparing the children to go to daycare.

In the morning after my husband left for work, I had to do things twice for the twins."

When the Russian honey bee project was initiated, the team had to evaluate hundreds of colonies in Iowa, Mississippi, Louisiana, and the laboratory colonies. "Spending at least four days in each location every month," de Guzman recalls, "members of the Russian team spent more time with each other on the road than with our spouses and children during the bee season."

"Behind all of my success was a loving and understanding spouse," she says. "I am very lucky to have a husband who supports everything I do. We have been married for 27 years. He never complained about me spending long hours at the lab or him taking care of the needs of the children while I was away."

In speaking of her career and the differences she sees between US labs and research in the Philippines, de Guzman states: "I respect most U.S. scientists/professors for their willingness to get their hands dirty. In contrast, Asian supervisors/professors are expected to stay in their offices and just wait for data to arrive. They may go in the field but only to briefly supervise. They are also proud to show their domination and power—that they are in control! Perhaps this approach has something to do with the availability of very cheap labor in Asia."

But de Guzman also admires the supervisor-student relationships in the United States: "There is definitely a line that separates the two but not to a point where students or subordinates are scared of supervisors which creates that sense of inferiority. I believe that friendly interaction between supervisor and subordinate is the best way to learn since learning is a two-way street."

De Guzman's status as a dual-minority scientist in the United States has given her an insider's perspective on affirmative action discussions. "Everyone has the right to be treated fairly without regard to the color of her/his skin, gender, age and religion," she asserts. "Affirmative action is a law and as good citizens, we should abide by it. But the reality is racism and prejudice will never become things of the past."

She continues, "I believe in working hard. I believe in earning what you get. . . . But the truth is, my gender worked negatively for me in a particular hiring situation." This particular experience had long-lasting consequences for de Guzman. In addition to the short-term blows to her self-esteem, the normal rewards that come with accomplishment are now tinged with suspicion: "I have developed fears—fears that are hard to overcome." But emphasizing how exciting it is that more women are going into bee-related research, de Guzman states that she learned some basic skills when she helped her parents with their farm in the Philippines: "Research requires a lot of creativity which can be reflected in the techniques and how resources are obtained and used. I do not believe in having to buy the most expensive materials or fancy equipment to be productive. If I can modify existing materials or equipment, I do so."

Gloria diGrandi-Hoffman. Another woman who learned to be creative is Gloria diGrandi-Hoffman, director of the Carl Hayden Bee Research Center in Tucson, Arizona. In her role, diGrandi-Hoffman fosters an environment in which everybody works as a team. Her team of researchers listens to beekeepers and develops a five-year plan to address their needs. Recently, beekeepers have been expressing concern about nutrition and pollen supplements. Since pollen is the basis of microbes, diGrandi-Hoffman's team has been revising its initial goals to fill in an overall vision of bee health that all federal bee labs contribute to. Sometimes the lack of technology has held back previous researchers, diGrandi-Hoffman explains, but ideally, all the federal bee labs integrate the information that each lab is learning.[42]

When asked about the difference between being a director and being a team member, diGrandi-Hoffman credits her background in organized sports to instilling respect for clarity in communication and treating everyone the same way. Being a first-generation Italian, diGrandi-Hoffman also credits her parents for teaching an appreciation for hard work and technology, and especially a joie de vivre. "My father woke up singing in Italian," she laughs. These qualities and core values translated very well to science and made the transition to being a director much easier. Another element that smoothed the transition was the delay in starting a family until diGrandi-Hoffman was in her forties. "The second happiest day of my life was when I saw the ultrasound. The happiest day was when she was born," she says.

UNIVERSITIES

Compared to the federal bee laboratories, North American universities have been more receptive to women bee scientists since the nineteenth century, but gender-based restrictions on admissions, salaries, and workplace protocol have meant that progress for women as faculty members or researchers has occurred in spurts. Women in the later twentieth century still contended with intermittent progress. Affirmative action laws in the 1970s brought admission policies at state universities in compliance with federal standards, so female students had more access to science departments, but salary equity for faculty remains a problem.

Nonetheless, the state university systems have provided women invaluable opportunities to change centuries of socialization. As women beekeepers have earned degrees and accepted positions in laboratories, extension service units, or classrooms, they have had greater responsibility in teaching, providing oversight for undergraduate and graduate students, and leading cooperatives. Representing a cross-section of university programs, the women interviewed for this section—Maryann Frazier, Marla Spivak, Susan Cobey, Jennifer Berry, Nancy Ostiguy, Diana Cox-Foster, May Berenbaum, and Christina Grozinger—have juggled graduate programs, committee work, and research with tenacity and grace.

Maryann Frazier. The nineteenth-century Hatch Act set up extension services for beekeepers, but the fledgling bee inspection service in the early twentieth century faced a number of challenges: poor roads, low salaries, and large geographic service regions. By the 1920s, the inspection service had merged with the extension realm of university

Maryann Frazier and Diana Cox-Foster, Pennsylvania, United States.
Courtesy of Steve Williams.

service. Until recently, bee inspection has been dominated by men because the profession could be hazardous, require long-distance travel, and demand heavy lifting. The 1930s were especially difficult times for bee inspectors, who often followed cattle inspectors. Since farmers and orchardists could barely afford basic care, many inspectors had to advise beekeepers to burn diseased colonies, which led to confrontations.

Maryann Frazier was a Maryland bee inspector from 1983 to 1985 and has been with Pennsylvania State University since 1988. A complex web of forces has changed bee inspections in the United States since the turn of the twentieth century. Technology shortens the distances that inspectors have to cover. Cars and four-wheel-drive vehicles make a huge difference compared to the horse and buggy. Some states have reduced the amount of terrain covered and fund county inspectors as opposed to having one person cover an entire region or state. The number of treatments available to beekeepers has increased for a variety of stresses, including varroa mites, chalkbrood, small hive beetle, and tracheal mites. "Fumigating equipment made a huge difference. People are not nearly as scared of losing hives," explains Frazier.[43]

Compared to the Great Depression, when beekeepers often had to burn their hives if the bees had foulbrood, fewer people now are so dependent on bees, so they are more likely to accept the bad news that hives need to be burned. "They are not going to lose their livelihoods," explains Frazier. More than 90 percent of beekeepers are small-scale apiarists, not commercial or sideline beekeepers. Many states, to encourage people to call an inspector, have stepped away from using inspectors as police, which was the position's reputation for a long time.

Still, there was an incident in 1984 when Frazier showed up a little early at a farm and started inspecting the farmer's beehives. Already in her veil, she heard his truck pull up. She heard his door slam. As he approached, his face turned florid and he started spouting expletives. In an effort to allay the situation, she began walking toward him, but only when she got out of the beeyard could she take off her veil.

The farmer had just enough time to launch into a full-blown rant about the encroaching power of government when he was cut short at the sight of the petite, auburn-haired Maryann coming from underneath the veil. In almost the same instant, his finger waving in her face, his cheeks burning with self-righteous indignation, he turned into one of the most courteous gentlemen she met on her circuit.

When working overseas from 1986 to 1987, Frazier found that being a woman was more of an asset than a hindrance in the Muslim country of Sudan and in the non-Muslim country of Uganda. Her biggest fear was that she would not be as effective as she was in the States. Instead, she found that as an educated Western woman, she was more effective because she could talk and work with both female and male scientists and beekeepers.

Now, she works as an extension specialist at Penn State, which means that she gets to do applied research on subjects such as oxalic acid and bear fences. She "loves being able to talk biology and management."

Frazier also is an effective advocate for pollinators. When dealing with colony collapse disorder, she took Congress to task for underfunding research, stating in her testimony to the Committee on Agriculture in the House of Representatives in 2007: "I believe the magnitude and timeliness of the response has not matched the scale and urgency needed to save an industry valued at more than $14 billion. . . . The funding that has been allocated to date falls far short of the time sensitive and potentially catastrophic nature of this problem." Perhaps the most important part of her testimony was a simple question that had no answer: "How would our government respond if one out of every three cows was dying?"[44]

Frazier follows in the footsteps of major women entomologists such as Edith Patch who urged more responsible use of insect controls: "We are becoming increasingly concerned that pesticides may affect bees at sub-lethal levels, not killing them outright, but rather impairing their behaviors and their abilities to fight off infections." In concluding her testimony to Congress, she explained, "There has been little effort invested in finding biologically-based alternatives to pesticides, including the most promising, the development of bees resistant to mites."

Although the funding for such biocontrols has not appeared, an important tool in managing honey bees with varroa mites has been to develop genetic resistance among honey bees in the United States. Women such as Marla Spivak and Susan Cobey have been leading genetic diversity research across the nation.

Marla Spivak. Not that Marla knew that she would be the herald for such a paradigm shift when, at the age of eighteen, she spent an all-nighter reading a book about honey bees. At that point, in the 1970s, African honey bees seemed to represent a more immediate threat to beekeepers compared to other threats at the time. But having read enough to know that honey bees were fascinating, in 1973 Spivak, along with her advisor at Prescott College, contacted Jerry Cole, a

Marla Spivak, Minnesota, United States. Courtesy of Dan Marshall.

commercial beekeeper in the Rio Grande. Prescott College, Arizona, was unusual because it allowed students to develop their own curriculum—and in many ways, Spivak has never stopped developing her own course. She states, "If there is discrimination against women, I don't know because I think I'm blind to it. Somehow I've never made a distinction between careers for men and careers for women. I've just followed my heart and have done what I like."[45]

So with a "that's that" approach, Spivak traveled to an international beekeeping conference in southern Brazil with beekeeping friend Abbie Zeltzer in the late 1970s. On a whim, they decided to travel across South America, through Paraguay and Bolivia and into Peru, meeting beekeepers. During this trek, Spivak became very ill, and while hospitalized, she met a doctor-beekeeper. The doctor had been losing his bees, and he encouraged her to stay and help him figure out the problem. This mystery led to other adventures summarized in the South America chapter.

Spivak's time in South America led to her research in Central America. In 1984 Spivak, with her three-year-old son in tow, moved to Costa Rica to study the African honey bee for her doctorate. Staying until 1986, she monitored the bees' movement through the country, focusing on whether they could adapt to colder temperatures and higher elevations. She learned that up to ten thousand feet, African honey bees adapted quite well. Even though there were nightly freezing temperatures, there were no sustained winters.[46]

Spivak happened to be doing a three-year postdoctoral degree in Tucson the same year that the African honey bee arrived in the United States. There was much cultural anxiety about the arrival of the bee, but Spivak had learned the importance of "taking things as they come" from her time in Latin America. She had a ground-level perspective of how African honey bees were changing the ecological landscape. As the African honey bees migrated north from South America to Central America, Spivak provided research that would enable North American researchers to dispel myths and theories.

Going from the tropics to the North American heartland, Spivak relocated to the University of Minnesota in 1993. Her research to develop a Minnesota hygienic line of queen bees had unintended successes. The result was a highly successful breeding line that has been able to provide genetic diversity for beekeepers in the United States. But the successful line of queens was more commercial than Spivak had anticipated. "The original intent was not [to] come up with a commercial line of bees, but rather to teach beekeepers to select for hygienic behavior among their own bees," she explained in an interview. In 2009, at the American Beekeeping Federation, she announced that she was following other research tangents, but for me, a person who became a beekeeper in the 1990s, Spivak will always be associated with the Minnesota hygienic queen because of its "designer" status. She, along with Susan Cobey and researchers at the USDA bee lab in Baton Rouge, clarified with never-wavering conviction that breeding and maintaining genetic diversity was the best way to control bee losses from varroa mites.[47]

Even if Spivak's only accomplishment had been to advocate for genetic selection and breeding practiced by ordinary beekeepers, she would have deserved our respect. The Minnesota hygienic queen came on the scene when beekeepers were dumping chemicals into the hives or producing nonresistant Italian queens.

Now, however, Spivak has turned her attention to an unglamorous by-product of the beehive—propolis, a sticky, mustardy substance collected from resins to form a protective caulk in the beehive. In the middle of summertime heat, propolis is a mess, getting on everything from clothes to fingernails to car fabric (when inner covers have been carelessly placed in the back seat). In colder temperatures, propolis turns into concrete.

Spivak thinks there could be a link between colonies that produce large amounts of propolis and hive health. Instead of seeing propolis as a necessary nuisance, Spivak thinks that it could be a visible clue of a colony's immunity. If a colony produces a lot of propolis, it may be better able to defend itself from pathogens. Spivak suspects that propolis, an antimicrobial "tool" of the bees, has been an overlooked factor in bee immune systems.[48]

Spivak is one of the participants in the Coordinated Agricultural Project (CAP). One of the objectives of this project is to research a method of understanding bee loss in nonmigratory beeyards. It also looks at pesticides, pathogens, and pest levels and is one of the first national studies of its kind. But one of the important preliminary findings of CAP is that queen supercedure (when a colony replaces a queen because she is inadequate) is a major problem for beekeepers, "something that deserves more research attention." Spivak states that the project, along with another grant to investigate the effects of different floral landscapes on bee nutrition and health, "has given me a renewed appreciation for the diversity, or lack of diversity, in the landscapes that our bees experience across the U.S."[49]

Fortunately, Spivak's contributions have not been lost on the wider world. In 2010 Spivak was awarded the prestigious MacArthur Fellowship for her creative solutions to honey bee challenges. The award is $500,000, with "no strings attached" so that fellows can apply their research with freedom. According to Kim Flottum in *Bee Culture*, Marla is the first honey bee researcher to receive the award.[50]

Susan Cobey. Joining Spivak in promoting genetic diversity has been Susan Cobey, currently at the University of California–Davis. I met Cobey when she was at Ohio State University during one of her popular instrumental insemination courses.

In 2005 Cobey, an unpretentious, slender blonde gliding around the Rothenbuhler bee lab in Ohio, met a bunch of beekeepers from all around the world with diverse interests from industry to research. A key part of Cobey's workshops is to stress that instrumental insemination is not as complicated a task as it sounds. It is simply a tool to control breeding; the real work is in the breeding and selection aspects. Cobey realizes that the problems associated with varroa mites are complex and the use of chemical controls will not be sustainable. Instrumental insemination is another option, but it starts with having drone semen selected for various traits from colonies that are good honey producers, gentle to work, and overwintered stocks.[51]

We were willing students. We understood the need for better honey bee stocks. We did not have to be convinced that genetic diversity is a good thing. The high losses of colonies worldwide have been an increasing concern for a long time. Cobey taught us in step-by-step detail the technique for instrumental insemination. Once we finished that class, I no longer had to be convinced that one queen can be worth $500 or even $700, although Cobey has continued to emphasize that the value is in the results of a long-term selection and breeding program.

"The challenge is that queens randomly mate in flight, and that makes controlled mating difficult" says Cobey. "[Queens] will mate with multiple drones, as many as 60, averaging ten to twenty, within a couple of days."[52] Because queens mate in flight, it is difficult to know how many drones they have mated, the quality of the drones in the drone congregation area, and the days they have had to mate. If weather is a factor, queens may not have adequate time to mate.

Plus, honey bees are sensitive to inbreeding, and if too many related drones are in the area, the resultant worker brood will be spotty.

Cobey started a commercial beekeeping business in 1981 with the goal of putting the Page-Laidlaw Closed Population Breeding Program into practice and developing a stock named New World Carniolans. These bees have a celebrated history of endurance. Dewey Caron, a colleague of Cobey's, best sums up her contribution: "Sue Cobey has been of immense service to the beekeeping industry," Caron says. "In developing the New World Carniolans, she has produced a bee suitable for commercial use as well as a bee that's ideal for our backyard beekeepers."[53]

Instrumental insemination is simply a beekeeper's tool, Cobey stresses. The point is to enable selection. As instrumental insemination is not widely used, improvements in the bee industry have been slow to be realized. We can select for specific traits, but this too often limits intracolony genetic diversity, so the colony is less productive.

So this is the dilemma that Cobey has faced—changing a mindset about queen production that has been a hundred years in the making. Beekeepers have become accustomed to buying relatively inexpensive queens with no consideration for resistance, temperament, concern for drone quality, or wintering capabilities. The queen production industry tries to meet that demand and those demands of migratory pollinators who need thousands of queens every spring to fulfill contracts. With beekeepers so focused on queens and queen availability, drone quality and availability tend to be overlooked factors.

Furthermore, the culture surrounding academic laboratories is in flux. Many faculty positions are not being renewed or replaced when members retire. In fact, in 2007 Cobey decided to return to California, resurrecting the Davis bee lab after the honey bee research program at Ohio State University was dissolved. That closing has caused concern about the future of the university bee labs across the country.

One readjustment that Cobey has had to make in California is to the seasonal differences between there and Ohio. "The California pollination season is much earlier, pushed by the high demands of the almond industry," Cobey says.[54] There is no winter in California, and winter was her best selection tool in Ohio. Untreated colonies either thrive or dwindle in early spring, making selection easy.

In February, breeder colonies must be made, and Cobey starts raising queens to establish the next test population to evaluate. By March and April, she is conducting instrumental insemination on a regular basis along with her classes. In June, weather conditions in the Davis area become hot and dry. "At that point, the worker bees kick the drones out of the hive," Cobey says.[55] She acknowledges that the New World Carniolan is a conservative bee and better adapted to colder climates farther north. Although the university once had many more hives, UC–Davis maintains approximately two hundred colonies, with Cobey working to increase that number as well as continuing her efforts to refine the New World Carniolan.

Jennifer Berry. Just as Spivak and Cobey have worked toward establishing genetic lines suitable for colder climates, another researcher and queen breeder, Jennifer Berry, has started working on a queen line that is adapted to her region in Georgia. One of the most articulate advocates for integrated pest management, Berry brings a cautious optimism to her role in the South, which is simultaneously adapting to African honey bees, a thriving commercial queen industry, and small hive beetle. Whether she is talking to a thousand beekeepers in a conference, gathering up stragglers to go into a hive in a beeyard, or just hanging out in a cabin, Berry is consistent in her message: North American beekeepers need to adopt integrated pest management in their practices.

Berry and I both backed into beekeeping, but she had more fun than I did. She started as a stage actress working a summer for the Shakespearian Festival in Dallas. She then became a stand-up comedian, joining a group called the Outpatients. When comedy did not work out, Berry bought a black truck without air conditioning and headed to Los Angeles to make it big. "I quickly, just barely, became a single-cell protea, that lived on the amoebas, that lived in the intestines of the very tiny minnows, that were eaten by larger fish, that

Jennifer Berry, Georgia, United States. Courtesy of University of
Georgia Bee Lab.

hadn't even become famous yet, in that ever so popular pond," she
carefully explains. One day, Berry met a waitress in a restaurant who
leaned over and said, "I was big, I tell you. Big." It was, Berry punctu-
ates, "a moment."[56]

Having a class with Keith Delaplane, a highly respected varroa
mite specialist, was another moment. Berry now works in Delaplane's
lab and advocates the University of Georgia's five main components
of integrated pest management: biology, genetics, culture, environment, and chemical treatments as a last resource.

Integrated Pest Management

In a paradigm shift from chemical treatments, beekeepers are encouraged to use five tools in the integrated pest management approach: biology, genetics, culture, environment, and chemical treatments as a last resource.

As a research associate at the University of Georgia, Berry manages the apiary and colonies—depending on the season and year, anywhere from a hundred to a couple of hundred hives. *Bee Culture* editor Kim Flottum explains that these hives "are used in all manner of research projects . . . from studying the effects of small cell foundation on varroa populations to pollination efficiency studies to the cumulative effects of a witch's brew of pesticides that may be found in beehives." But Berry is also an entrepreneur, developing her own line of queens from survivor stocks of bees. "Looks don't make a queen in this operation," Flottum states bluntly. "It's her performance that counts, first, last and in between."[57]

From the time a graft is put into a queen cup, to the time the cell in which the graft is made develops, to the time the queen is eventually shipped via the post office, a queen bee is evaluated in Berry's operation. The queen gets checked to make sure she has mated, then that she is laying, and then that the brood pattern is solid. "During this inspection period, if there's a reason to remove [the queen] from the colony there's no hesitation to do so . . . no mercy for a queen that lays poorly, or just isn't cutting it."[58]

The queens are a mix of Italian, Russian, Carniolan, and hygienic genetics. Bees exhibit hygienic behavior by grooming each other, by removing intrusive material more quickly, or being proactive if they sense brood trouble and puncturing the brood cells to remove the larvae. Hygienic behavior is a genetic trait, and so beekeepers can select for this quality. Because of the variety of queen bees, the varroa counts in Berry's hives are low. Since Berry does not use chemicals, she dusts her nucs with powdered sugar in the spring and again in the fall.

Although the long-term goal is to have pure survivor stock, the short-term goal is to deliver high-quality, gentle queens that have low mite counts without having to use chemicals. Ensuring the queens have good drones to mate with is part of Berry's program. The "drone mothers" are as hygienic as the queens she intends to sell. Evaluating the brood patterns is part of ensuring beekeepers get the quality they pay for: if a beekeeper makes a conscious choice not to use chemicals, then a queen who is a good layer is a necessity. When dealing with varroa, numbers count. Making sure that the bees exhibit hygienic behavior is another part. Honey production is on the list, but it is not as high a priority as hygienic behavior.[59]

So, in Berry's queen production business, nothing is left to chance. Despite her easy casualness in the beeyard, Berry is changing the beekeeping culture by addressing the issue of poorly mated queens. In part, this issue can be blamed on environment and the weather, but it is also a result of human error, of people not being selective enough in the evaluation process or relying too much on one tool in the beekeeper's box. This is where Berry changes the way business is done. By keeping her business small, she can have more control over all the factors, and integrated pest management seems to lead her to new genetics as a key resolution to problems with queens.

Nancy Ostiguy. Bringing a new tool to the discussion of integrated pest management, Nancy Ostiguy uses her background in public health to show similarities between human societies and honey bee colonies. After a career in public health, Ostiguy graduated from the University of California–Berkeley in 1982 with a master's in public health. She says, "The parallels between honey bee society and human society only became more obvious as I learned more about honey bees. The social behavior of people and bees, as they pertain to the transmission of pathogens and probability of disease susceptibility, are incredibly similar."[60]

Ostiguy's research teaches beekeepers how to prevent problems with colonies based on how people prevent problems in their own living environments. For instance, if we compare a colony to a tenement or an apiary to a crowded city from the nineteenth century, we can see the correlation between space, clean water, and food resources. If we look at overcrowded public housing, we are likely to find higher rates of crime as the availability of resources is reduced, higher rates of disease because people have closer contact with each other, and higher rates of stress as social boundaries are crossed.

These same kinds of issues affect beeyards, Ostiguy argues. The lack of proper spacing will lead to poor distribution of resources—in this case, pollen and nectar—and perhaps the robbing of hives. Tighter quarters can make it easier for mites and viruses to spread. Clean water sources are also important because bees are exceptionally clean creatures. Ostiguy also recommends not moving bees or honey between colonies because such movements, as they would in a city, transfer bacteria.[61]

Regional differences affect varroa, although there is generally uniformity within a hive because bees control humidity. Beekeeping is not a "one size fits all" activity, Ostiguy sums up. This applied ecological method offers beekeepers an external approach to varroa, monitoring the environment as well as internal activity within the hive.[62]

When asked her opinion regarding gender in the sciences, Ostiguy provides this refreshing perspective: "I come to beekeeping with two parts of me being an outsider. My gender is one but so is my academic training. Both of these give me 'permission' to think outside the box. I can suggest more 'touchy feely' ideas, e.g., bees don't 'like' to be put on the back of a semi and be driven several thousand miles. Men might have to provide hard facts to have people even consider this idea."[63]

"The only other advantage," she continues, "if it can be called an advantage, to being a female in science is that you are noticed because you are not 'one of the boys.' . . . My academic training is in Environmental Toxicology. This has been advantageous when discussing the problems associated with miticides. My training has [also] been a disadvantage because there are some that don't believe I will ever know enough about bees."

Diana Cox-Foster. Diana Cox-Foster approaches the challenges of honey bee collapse by studying host-pathogen interactions within the hive. After receiving degrees at Colorado State University and the University of Illinois at Urbana-Champaign, Cox-Foster moved to Pennsylvania State University, where she is a professor of entomology and insect biochemistry. "I am interested in the coevolution of insects with their pathogens and parasites and the role the insect immune system plays in this interaction," she says simply. But there is nothing simple about the interactions that can happen in host-pathogen relationships. She began to work with bees in the late 1990s, looking at the interactions of honey bees, varroa mites, and viruses. "It was the interplay of the three that fascinated me," she explains. "Having [such a] diverse background has allowed me to develop interdisciplinary teams and to ask questions ranging from what happens to bees and other pollinators in the environment, to looking at mechanisms at the cellular and molecular level. My background and experiences in agriculture [have] given me the basis to understand how important pollinators are to our food production and natural ecosystems."[64]

Since varroa mites continue to be a major factor affecting honey bees, Cox-Foster's research applied a multilayered approach to the interactions between honey bees and varroa mites, considering multiple consequences of the host-pathogen relationship. She began to approach varroa mite–honey bee interactions in this way beginning in 2005, "when my student Xiaolong Yank found that varroa are inhibiting the immune responses of the bees and that bacterial exposure can have a role in virus infections." Cox-Foster continues with the explanation: "These viruses can infect bees [by] a myriad of different transmission routes. Once a colony is infected, the bees may exhibit little symptoms until stressed, such as by varroa mite parasitization and other factors."

From a beekeeper's standpoint, Cox-Foster's research is important because she is learning how stress factors such as the mites, pesticides, and lack of good nutrition can affect these viral infections. Many of the viruses may invade the bee's nervous systems or other essential organs and have major impacts on the bee and the colony as a whole. Minimizing these stresses and the spread of viruses is one of the goals of Cox-Foster's research. So, Cox-Foster focuses on clarifying the mechanisms underlying immune responses in insects and disease regulation. Using a chemical ecology approach, she hopes, will lead to better health of honey bees and other pollinators. This ultimately may lead to a better environment for her daughter. In this sense, Cox-Foster follows in a long line of women immersed in honey bees and sciences. Her great-great-grandmother moved to Colorado to open a bakery in a gold-rush town. Her great-grandmother was a beekeeper of high regard in Colorado, and her grandmother majored in food science and chemistry at Colorado State University. "Being a scientist and asking questions about the world around us can be great fun," she sums up. "There is still much to learn about bees and their interactions with the world. I can't think of anything better to do."

May Berenbaum. University of Illinois entomology professor and department chair May Berenbaum already had a distinguished career before the crisis known as colony collapse disorder, an event that made media headlines in 2006. The severity of bee losses around the world prompted nations to consider their agricultural programs. Unfortunately, quick solutions to the bee losses have not been forthcoming. However, researchers have been making headway. In 2009 May Berenbaum and her graduate student Reed Johnson were "not able to find a smoking gun, but did find a bullet hole." Using new molecular methods in which colony collapse disorder bees and pathogens were studied side by side, Berenbaum and her student theorized that the RNA in the honey bee colonies affected by colony collapse were damaged. In effect, when the honey bees can no longer process protein, the hive will be susceptible to any number of problems—for example, Israeli acute bee paralysis virus, deformed wing virus, and nosemae. When asked if it was like HIV in humans, Berenbaum answered affirmatively, with a major qualification: "Instead of immune systems being attacked, the bees' protein systems are being attacked." Berenbaum was careful to emphasize that there is much work to be done with colony collapse, but for her efforts, Berenbaum was awarded the prestigious Tyler Prize for Environmental Achievement in 2011.[65]

Christina Grozinger. Approaching hive decline from the topic of queen pheromones is Christina Grozinger, who was at North Carolina State University and is now at Pennsylvania State University. She brings a molecular biology approach to honey bee behavior. After finishing her doctorate in chemistry and chemical biology at Harvard, she was searching for a larger research trajectory. Previously, she had focused on neurobiology and behavior among fruit flies because they are easy to manipulate.[66]

As it happened, her brother had become a beekeeper and enticed Christina to think about honey bees. "I was just fascinated by the facts he would share," she explains. So, Grozinger began to seek a lab to do further research. She landed a postdoc position in the lab of Gene Robinson, a professor of integrative biology, at the University of Illinois Beckman Institute. Robinson is not just any professor. He is world-renowned for his leadership in mapping the honey bee genome sequence. Under his tutelage, Grozinger began a career that focused on chemical communication among honey bees. Robinson explains: "Christina made a huge switch in research paths by joining my lab. She had never before worked with a whole organism!"[67] Even though Robinson is more interested in foraging and honey bee behavior than she is, Grozinger found in Robinson a great role model and mentor.

For someone who likes mechanistic models, Grozinger seems to be equally compelled by the mysteries of chemical communication. Pheromones within the colony are passed between queen and workers, workers and workers, and queens and drones. One pheromone in particular, the queen pheromone, modulates the colony activity. Before Grozinger and others began turning their attention to queen pheromone, most researchers had assumed that the pheromone was relatively static. The queen was either mated or not, and the bees would respond. However, when manipulating pheromones to see their effects on honey bees, Grozinger would be puzzled by the fact that bees would not do as they were expected to do. "Why are there so many variations and responses?" she laughs with frustration. "It seemed that there is a lot of variation in queen pheromone."[68]

In working with David Tarpy in North Carolina, Grozinger has expanded some theories regarding queen pheromone quality. Tarpy theorized that the number of matings a queen has affects the quality of her pheromone. Grozinger expanded that theory by considering other factors such as nutrition, pesticides, mating frequency, and viruses. If these external factors affect the queen, what and how then do the workers react to her? Is this one reason why queens are superseded?

"I thought there was something to be learned by looking at how mating number affected queen pheromone and queen-worker interactions," she explains to me. "At the University of Illinois, we did a lot of work with single-drone inseminated queens because we try to use workers that are highly related for many of our experiments. The workers try to supercede them [the queens] all the time—we need to go through the colonies at least once a week to remove queen cells. I thought that it could be because the queens were 'poorly' mated, with semen from only one drone."

Grozinger goes on: "When I moved to North Carolina State University, I was excited to collaborate with David Tarpy on this project because of his expertise in queen behavior and ability to do instrumental inseminations. But the project was funded by a USDA grant that I wrote with David—I was the PI [principal investigator] and he was the co-PI—and it was conducted by a graduate student, postdoc, and technician from my lab. We worked closely with David for setting up the field experiments, but the rest of the work—the chemical analysis, genomic studies, and physiological studies—used expertise from my lab group."

Grozinger's baby, Evelyn, interrupts this discussion with a just-awakened-from-a-nap grumble for attention. "Hold on for a minute," Grozinger stops. "I just need to hand Evie off to my husband."

Then Grozinger and I return to talking about the politics of research and academe. "I haven't been aware of discrimination in a formal manner in my career," Grozinger explains. "But sometimes I wonder if it is a matter of how well one sits at the table. If you share a lot of interests and conversations with male colleagues, then inevitably you get included in a lot more opportunities."

Perhaps another factor has been the mobility that has come with having two parents as scientists. As with other women beekeepers in North America, Grozinger has hopscotched all over the place. Her parents were from Germany, and they immigrated to Canada, where she was born. When she was three, they moved to Connecticut. Grozinger always felt that science was a way to understand the world. Chemistry, especially, "offered mechanistic models for things that didn't make sense in biology," she offers. So she went to school at Harvard, moved to Illinois, and then accepted positions in North Carolina and Pennsylvania.

But some issues remain the same no matter where a woman goes. While in North Carolina, Grozinger led a discussion group focused on women's issues in entomology. She has been very much aware that institutional hurdles exist, and the group would concentrate on channeling frustration and distinguishing between personal problems and institutional problems. "It was important that the group understand the underlying causes and be aware of that, and not take the institution's problems upon themselves," she explains. "My proudest moment was when a new participant showed up, and she expressed concern about balancing work and child care. The rest of the group immediately began explaining the institutional options, and how having access to affordable child care on campus would go a long way towards solving some of these issues."

When asked about institutional issues that departments face when selecting candidates for interviews, Grozinger replies thoughtfully, "A more central concern among entomology departments, especially those with a big focus on extension, is who can talk with the growers, who are assumed to be men. At times, it seems that the logical answer to the question precludes certain types of people. At times, the hiring of new faculty might be male-biased, particularly in positions involving a lot of interaction with farmers, beekeepers, and other stakeholders, because there is a perception that men will relate better to predominantly male stakeholders. But, that really limits the choice of candidates and the best-qualified person may be missed."

Grozinger is now director of the Center of Pollinator Research, which capitalizes on the many honey bee researchers and extension people at Penn State. "We just decided to formalize the network," she explains. "We're trying to create opportunities, create more positions, and provide training grants for students." It is also a way to systematize an approach from basic to applied research. But she laughs off the notion that her position as director is in any way tied to a hierarchy: "They tell me what they want, and I make it happen!"

QUEEN PRODUCTION AND LABOR

Although the federal labs and university programs provide education and research to the commercial bee industry and beekeepers, two more factors are as important to the industrial apiculture system in North America: unique land conditions and an available labor force.

In the commercial queen production industry, no two places in North America are better suited for production than California and the Big Island, Hawaii. These two states have semitemperate and tropical climates as well as nomadic and eclectic laborers willing to work for minimum wage. The queen production schedules are closely tied to weather events and floral patterns. Depending on the queen production business, most companies have at least three multifaceted crews working together—a queen-catching crew, a grafting crew, and an office staff managing the paperwork— although depending on the size of the operation and the number of employees, there can be other crews focused on honey production, breeder queen yards, truck fleet maintenance, and so on. Whereas a honey producer may have problems with the hives and choose to use a chemical treatment to treat the bees, queen producers have to find other alternatives if possible.

Jackie Park-Burris. Queen producer Jackie Park-Burris states simply, "There can be no short cuts when it comes to queen production."[69] Taking over her father's queen business, Park-Burris has been producing Homer Park Italian queens since the 1980s. Park-Burris's uncle, Homer Park, started keeping breeding bees in the 1940s, and soon her father became a beekeeper too. For forty years, the Park-Burris family exported queens to Canada, until the borders closed on January 1, 1988, because of mites. The transition was difficult. "That date is burned in my mind," Park-Burris explains, "because it is like having a retail store and being told Christmas is cancelled." Not only that, but her father was diagnosed with a serious illness.

Park-Burris credits their survival of those challenges to the quality queens they produce. A major factor in her business is having an ample supply of drones so that virgin queens have plenty of opportunities to mate. "These queens are so prolific," she says. "Compared to other genetic lines, Homer Park Italian queens will keep laying eggs through adverse conditions and are good honey producers."

"They are also prettier than other queens, too," she proudly points out.

It is not easy getting those pretty queens. "You really have to have your act together and have a little help from Mother Nature," Park-Burris acknowledges. When asked about being a woman in a male-dominated industry, she responds: "You don't cry or whimper when you take stings. You take it with a smile."

That determination makes it easier to deal with larger challenges. Many times "a woman has to have attitude," Park-Burris asserts. In general, commercial beekeepers get territorial when another beekeeper has moved hives too close to theirs. Honey producers do not like their bees having to compete for floral resources, nor do they want their bees at risk for contagious diseases or pathogens. But that sensitivity is increased for queen producers, who have to maintain genetic stock of not just queens but drones. "With queen breeders, the drones are as important as the queens. For every queen, you have to have twenty good drones," Park-Burris explains. It is not unusual among male beekeepers to protect company mating yards with a bare-knuckled brawl.

That is not an option for Park-Burris. She explains she "has to be creative" when communicating with her colleagues who may move bees too close to hers. "I can't beat them up, so I have to talk through problems. . . . It would be so much easier to have a fight and go on but I cannot do that."

The Park family enterprise also has been a huge source of support, although they have four separate and distinct businesses. While Glenda Wooten handles the marketing of Homer Park Italian queens for all four family businesses, Jackie maintains her own queens and nucleus hive business. Yet all the trucks' CBs are dialed to the same number. If a family member gets stuck in a young orchard, someone will be there to pull the other out: "That's how it works." Park-Burris keeps things simple: her truck fleet has one-ton boom trucks, which she inherited from her father. Even though most queen breeders have gone to putting nucs on pallets and moving the pallets with fork-lifts, those boom trucks come in handy when dealing with her larger nucleus hives.

By far the biggest biological challenge to the queen production business continues to be varroa mites, because "it is hard to find a window of opportunity to use treatments in California," explains Park-Burris. Other challenges are the weather, which nobody can do anything about, and Australian packages, which are often loaded with small hive beetles. When California has shortages of pollination bees for almonds, such as in spring 2010, migratory beekeepers will add Australian packages into weak beehives. Park-Burris and other California bee breeders are working in a united way to develop a national protocol for package importers to make all imported bees safe for the United States. When asked if a national protocol is feasible, Park-Burris explains that there are already protocols in place for other types of honey bees. Researcher Tom Rinderer has to register Russian queens for the Russian honey bee project in Baton Rouge. Susan Cobey has to have the drone semen she brings from

Europe isolated on an island in Washington State. The same kinds of protective measures need to be put in place for the domestic honey bee industry, and Park-Burris is optimistic that this standard can be achieved if people work together to make it happen.

Valeri Strachan-Severson. Valeri Strachan-Severson is another major queen producer in California, inheriting her business from her father, Don. Valeri began working with her father in 1975: "We worked side by side until he retired in 2001 and died in 2003. It was a fairly easy transition for me to follow in his footsteps." But there were personality differences. Strachan-Severson explains: "My dad was more of a commander-in-chief. I tend to be more of a delegator. I like to think I can depend on others to do their job as a supervisor or crew boss. But I'm still the one who they ultimately answer to. It's not unlike being a mom."[70]

Strachan-Severson offers specific examples of the differences in her dad's approach. "I would often challenge him in things he wasn't comfortable with. These things included working with computers in the early nineties to taking classes in instrumental insemination or driving trucks! He would eventually allow change to take place, and in time he could see the benefits. Most of the challenge in business is like this . . . it's just viewed from a different window."

Now that Strachan-Severson has taken over the business, she works with Susan Cobey to offer New World Carniolans. "I'm also the one who maintains our line of New World Carniolan stock through instrumental insemination," she says. "I have a couple of employees that help me collect drones and also prepare the queens for this task. After our queen season slows, I spend two weeks or more instrumentally inseminating breeder queens. It is the best way to maintain our superior stock. It is also the best way to control the desired traits you want in your hives." She often works with Sue Cobey to introduce new blood lines. "We've had a good working relationship for a long time," Strachan-Severson says. "She and I share the same opinion of the stock we work with and care about. Occasionally, I'll experiment with other lines of bees to see if there's a way to improve what we've developed. That's the fun part of what I do."

Queen production businesses depend on successful queen grafts and catches, which can sometimes mean employing two different crews. Since M. H. Mendleson began hiring women as grafters and catchers in 1919, California has a history of hiring women for these roles. Strachan-Severson is no exception: "Many queen-raising operations now have women grafting and caging the queens. On my crews, I have a team of two men and two women grafting and caring for the cell builders. The work is demanding and tedious. It is true most women are a bit more careful with delicate work, but a few men are as well. If you get a good team together, it works very well."

She goes on: "My queen-caging crew is a team of five women and two men, and that has worked well for my business. On both teams the men are also year-round beekeepers, so their knowledge of bees is useful when it comes to recognizing any problems that arise while working with the queen bees."

Specific qualities that Strachan-Severson thinks have been helpful as a queen production business owner have been time and determination. "Over time I have earned the respect that comes with hard work and success," she says. "I don't think it's any different for a son, except it may take a little more time for a woman to develop the trust among an industry that is male dominated. The biggest asset for any woman in developing a business is determination."

Strachan-Severson also clarifies that communication is key, especially in queen rearing: "It may be easier now than when I started because there are more women in beekeeping. Since we are in a field where the majority is male, women have to develop a thick skin and not be too sensitive or you'll fail. Understanding how to communicate—and I'm not the best—with men as well as women is very important." Her biggest challenge, she says, has been "how to communicate on an equal level of understanding." Nevertheless, she asserts: "I don't think things need to change necessarily. We've already become a more accepting society of all genders. It's the job you do that is the proof. Queen production is like working with dairy cows. It's a seven-day-a-week job, although it is not year-round. Our queen production crew works every day during the months of March through May."

Big Island Queens. When varroa mites devastated the queen production industry on the mainland in the 1980s, the Big Island, Hawaii, became a mecca for queen producers. Until recently, Hawaii had strict import laws, and with its tropical climate, it was a good place to develop the queen industry. Varroa and tracheal mites were nonexistent there until 2006. African honey bees are not present as I write. Small hive beetles did not appear until 2010. Hawaii can rear queens throughout the year, and when commercial queen producers need to requeen in spring or fall, Hawaii is a reliable place to turn for good-quality queens.

In 2006 I decided to work for Big Island Queens in an effort to learn queen production. Queen production is a difficult topic, and I thought that being in the field would make certain processes easier to understand. The owners of Big Island Queens, Pam and Randle Brashear, had been beekeepers in Texas before selling their business and moving to the Big Island in the 1980s.

Being in queen production is not easy. Someone has to answer the phones, which ring at 3:00 a.m. because mainland people forget about the time differences. Pam's part of the business was dealing with the paperwork, answering phones, maintaining Internet service, and handling postal orders. Postal service shipping rules changed arbitrarily, leaving Pam to drive queens to the post office only to be told that it was impossible to ship queens to the mainland. The next day, the same package would be shipped without question. Queens would also sit on the tarmac for days if the shipping rules changed again after she had dropped them off. Pam's motto for the Big Island Queens business was simple: make plans, but do not plan the results.

For much of the 1990s and the first decade of this century, the Big Island Queens catching crew was led by Joanne Murray, who

Joanne Murray, catching queens, Hawaii, United States. Courtesy of Perry Amos.

hailed from New Jersey but came to the Big Island and never left. In the year I started catching queens, the crew was a fluid bunch of people, with some workers in Hawaii for a couple of months, others only for a week. The other crew members were Hunter, from Virginia, who was working in Hawaii for a few months; Daniel, who eventually moved to California; and Joe, whose girlfriend was pregnant with her first child.

On a typical catch day, the crew catches queens with the thumb and forefinger. No gloves. Crews often work without a smoker or a veil.

Each queen catcher has idiosyncratic tools. Most have cigarette lighters in case they need to light a smoker. Some have spray bottles to clean their hands after work. Queen catchers will use a stool to sit by a nuc, open it, and pull out a frame. Some of these catching stools are works of art. Colorful names, hearts, and other identifiers dress up the stools in addition to the pastel-colored nucs. One crew supervisor would carry a machete to chop coconuts or avocados during breaks. I generally had factor-50 sunscreen.

Whenever a new queen catcher is hired, he or she is paired with an experienced catcher. My first day, Joanne and I sat opposite a nuc, without gloves or smokers. Mangos and avocados hung from nearby trees. One hundred and fifty nucs were positioned like Easter eggs in serpentine rows, and bees flew everywhere.

We opened the top of a three-frame nuc. "You want to start in the middle of a super; she generally tends to be there where the brood is," Joanne explained, taking a frame out very slowly, not rolling any bees. "You also do not want to use much smoke, which forces her to seek out the corners."

Glancing over the frame quickly, she saw the queen waddling off the side. "There she is! See her color! Your eyes catch that quick movement. Remember, pick her up by her wings!" Joanne encouraged, but I just wanted to sit and stare at this queen pacing the frame.

Catching a queen takes some practice. The wings seem like gossamer, but they are stronger than their appearance. It is easier and gentler to scoop the wings from behind the queen's abdomen. The one place you never want to catch her is by the abdomen, where the eggs are stored.

Joanne can out-catch almost everyone. Only field manager Perry Amos, who has been catching queens for ten years, out-catches Joanne on a regular basis. In my third season of being with Big Island Queens, I am a solid middle-of-the-line catcher. I'm not the slowest, I'm not the fastest. Every now and then, I have my day when I catch the first box of queens before everyone else. But the others will catch more quickly.

There is an easy camaraderie that develops when moving nuc by nuc. We look like turtles. We work a row in step-by-step fashion. One will inevitably get hung up on a queen who has become so elusive that the catcher will have to look at frames repeatedly, close the nuc, and go back and look again. At times, the queen walks onto the lid when the catcher is focused on the frames or in the bottom corners. Sometimes a queen will take it upon herself to just fly away from the whole scene, and then the chase really begins. Good catchers will reach for her in the air. Other times, she will fly under the hive. Every now and then, you'll hear someone yell jubilantly, "Top bar!" and know that the catcher just opened the nuc and the queen was on the top of the frame, no looking required.

After the catch is done, the crew will finish by placing queen cells into the cages. The cells are approximately ten days old, so when the queen emerges by the eleventh day, the workers will be ready to accept her. She will need eight days to mate and four days to lay eggs. Then the catching process begins again.

When catching queens, the crew discusses topics running the whole range from politics to risqué jokes to relationships in their various stages. Working like that, long day after long day, you not only learn the bees, you learn each other. Tempers can get short at the end of the day.

The graft crew requires a different rhythm. The grafting process starts outside, by selecting hives that have certain traits a beekeeper wants to reproduce—for example, gentleness, good honey production, or hygienic behavior. Or perhaps a beekeeper wants bees that have good numbers but are not inclined to swarm, so we would choose larvae from those bees. Others choose to graft from hives that have done well in spite of higher mite counts, in an effort to breed genetic resistance. Sometimes, commercial queen yards bring in other strains of honey bees—for example, queens with varroa resistance—in an effort to have some genetic diversity.

For most conventional grafting operations, a beekeeper needs good eyesight and good eye-hand coordination. Warm, humid rooms are a must, because larvae are sensitive to light and dry conditions. A grafting room needs to mirror the internal conditions of a hive.

In my opinion, grafting is like doing needlepoint in hell.

The calendars in queen grafting rooms have to be accurate, or the queen-catching crew and the grafting crew will be off schedule. The best queens are grafted from larvae on the fourth day, curled into a "C" shape on their side. At times, it is easier to just aim for the royal jelly. Once the larvae are picked up on the grafting tool, they are placed into a queen cup and put back into breeder colonies for the workers to rear into queens. The breeder colonies have to be

Commercial Queen Producer Schedules

Commercial queen producers are always on two schedules at the height of queen production season:

- A ten-day schedule from the time the queen graft is placed in a hive to the time of cell maturity.
- A sixteen-day schedule from the time the queen emerges from the cell. The queen will mate on the eighth to twelfth day after she has emerged, and she needs to lay at least four days before she is caught. (This schedule is highly variable among commercial queen producers and depends on weather conditions.)

jam-packed with healthy bees and fed a consistent supply of pollen throughout the process.

Queen grafters seem to be predominantly female. Big Island Queens office manager Klarene Olivarez would demonstrate how easily her daughter Krissy could transfer larvae to a queen cup in one smooth motion. "Just like this," Klarene would say, her wrist moving smoothly in one motion from the frame to the wax cup. Much more methodical, almost as if performing scientific analysis, Joanne Murray has been grafting since she moved to the Big Island in the 1980s. For her, grafting queens has become a necessary skill.

As with other agricultural industries, queen production is subject to variations in weather. Inevitably, there are die-offs in the queen industry, times that the grafts do not "take" or environmental conditions change and the queen cells are not accepted by the worker bees. Sometimes, virgin queens will leave the hive to mate and not return, leaving queenless colonies. Even when the queens are not doing well, I learned how vital women workers were to the queen industry.

MIGRATORY POLLINATORS

The most important factor missing from Steven Stoll's discussion of the industrial agriculture framework in North America is the migratory pollinator. This became apparent in the 1980s. North American hives had no resistance to the one-two punch that varroa and tracheal mites caused. As hives collapsed, pollination services became even more valuable, and they continue to be more profitable than other bee-related industries. The 2000 census showed that pollination contributed $14 billion to agriculture, and it is safe to assume that number is much higher now. The United States still has trouble maintaining enough colonies to provide pollination.

One answer to this North American crisis has been to import honey bees from Australia. In 2004 Congress decided to relax the 1922 honey bee import ban so that California growers could import bees from Australia and thus avert an almond crop disaster. But in 2010 California was two hundred thousand hives short of its pollination needs. As recently as 2006, commercial beekeepers had to go through arbitrary inspection rules. Border inspectors, at times, have not known the laws when it comes to other insects that may have wandered onto bee trucks. Fire ants have been an issue. Trucks of bees have been sent back to Texas after waiting to be inspected because an inspector saw one fire ant. Sometimes inspectors have not been at the station. Some states have to inspect bees before beekeepers leave their home state; other beekeepers have no inspection until they reach California. Because each state is different and each agricultural year is different, the beekeepers are put in flux. Two migratory women beekeepers, representing opposite ends of the country, bring different perspectives to this invisible industry: Bonnie Woodworth of North Dakota and Stephanie Tarwater of Tennessee and Florida.

Bonnie Woodworth. Bonnie Woodworth has followed the arbitrary nature of the commercial bee industry since the 1970s. She married her husband, Brent, in 1972 and began migratory beekeeping in 1973. They took their bees to Texas in the winters. She and Brent struggled through the 1980s with many years of drought in North Dakota. She calls those years the worst of her life.[71]

In 1991 the Woodworths started migrating to California because they could not survive any other way. Pollinating almonds in California was much different than wintering bees in Texas. It was a new challenge and a total lifestyle change, but a welcome one.

Woodworth continues to provide leadership and to be a voice for the pollination industry. But she is realistic about the difficulties of the industry. It has become necessary to hire workers from South Africa rather than local people because of the lack of available workers. Honey production can be tedious. And with the recession affecting farmers' and orchardists' ability to care for their crops and lands, migratory beekeepers such as Woodworth will continue to feel the economic effects long after the more visible industrial agricultural framework has recovered.

Stephanie Tarwater. Stephanie Tarwater grew up scrapping hives on her grandfather E. R. Tarwater's porch in Tennessee. She became a migratory beekeeper after graduating from Pellisippi State Community College, Knoxville, in 1999. In 2002 she finished an advanced degree in agriculture at the University of Tennessee. In the late 1990s Stephanie ran bees between Georgia, Tennessee, and North Carolina. Her bees pollinated winter vegetables such as yellow squash, pumpkins in Tennessee, cantaloupes in Georgia, and watermelons in North Carolina. She worked a lot with Holly Jones, a Nashville honey producer. As with many of the migratory beekeepers, Tarwater developed strong repeat-business clients.

When asked if being a female was a disadvantage in the migratory business in 2005, Tarwater shrugs. "I've been in the business a long time," she says casually, her signature ponytail motionless for a moment. She has grown up working bees with her father, Joe Tarwater, and her mother, Sue. Her entire family goes to bee conferences, and her social networks are deep. "Have there been problems? Yes," she acknowledges thoughtfully. "But if there is a problem with pollinating a crop, if the field isn't right or the guy tries to renege on our agreement, that's the last time I'll bring bees to that man's yards."[72]

Three years ago, Tarwater became an inspector in Florida and Tennessee. It was a way to have a more secure income with benefits and health insurance. Since Florida's contracts are not for the entire

Stephanie Tarwater, Tennessee, United States. Earning three service awards for her work as a bee inspector, Tarwater poses with her colleagues Jerry Hayes and David Barnes. Courtesy of Marianna Beckman, Florida Division of Plant Industry.

year, she returns to Tennessee to run bees. But being an inspector is not a major shift. Both her grandfather and father served as state and county inspectors after having careers at Alcoa. "Papaw always had a hundred hives around and always gave away more honey than he sold," she says.[73]

Even though she's a fourth-generation beekeeper, Tarwater still sees people do a double-take when she shows up to inspect hives. "People still say, you're awfully young and being a woman, well, the men insist on moving supers. But once they hear me talk, and figure out that I am who I say I am, it's never a problem."

This flexibility between serving as inspector and running bees seems to work for Tarwater. In 2006 she became an award-winning inspector in Florida. She received a certificate of completion for training on FABIS (Fast African Bee Identification System), the quick "field test" inspectors can use to check bee wing vein morphometics as a screening tool to determine a colony's subspecies. Tarwater also completed training for the more specialized morphometic identification for African honey bees, USDA-ID (Universal System for Detection of

African honey bee Identification). A special recognition award honored her willingness to go above and beyond her inspection schedule. None of those awards are surprising, frankly, for when Tarwater talks about the need for greater cooperation among state officials, she means business. Since Stephanie works in the field and with the two different states, she walks a fine line between knowing what can be done in a realistic manner and recognizing typical bureaucracy for the hindrance that it can be. It is a similarly fine line when it comes to moving bees. Tennessee has one inspector for the whole state, and the law requires that migratory bees be inspected every six weeks. Scheduling appointments with county inspectors is difficult because they have other careers. Tarwater can move bees within a week, so trying to stay within the law is difficult.

Florida, in contrast, has thirteen inspectors as well as a more secure network between university research and extension personnel. The state inspectors go through regular training updates and are accessible to anyone who calls.

Tarwater appreciates the fact that in beekeeping, "there are lots of degrees to people's businesses. Some people can just do pollination and be happy. Some people do honey, honey, honey . . . that's their mindset." Other beekeepers may not see eye to eye with honey producers. Being both inspector and beekeeper, Tarwater can see how difficult it is to get legal and stay legal.

Migratory beekeeping may not be a feasible occupation for all women, but both Woodworth and Tarwater demonstrate that it can be a family affair, a stepping-stone to another career, or an unconventional supplemental line of income. Tarwater cautions, however, "It is easier for family because women have to be flexible. It is not just a one-person gig."[74]

Innovative Categories

In North America, women beekeepers create new ways of approaching the bee business. This chapter's conclusion emphasizes many categories in which women beekeepers have carved out unconventional careers, such as swarm removal, pollination management, specialized extension roles, and bee cooperatives. This chapter's concluding categories are not meant to be definitive but rather suggestive of the true range of possibilities, which is beyond this book's capacity.

CINDY BEE: SWARM REMOVAL SPECIALIST

Following a long line of teachers-turned-beekeepers, Cindy Bee started teaching English in Colorado. But after teaching high school for a while, she says, "I found the pay was so terrible, I could barely afford my student loans." Cindy got into bees full time after she realized the need for qualified swarm removal specialists. It didn't hurt that swarm removal paid much better than teaching. "Unfortunately or fortunately," muses Cindy, "the removal business has paid so well that I'm having a very hard time getting out of it. I love the bees and get a kick out of removals, but what I long to do is write. I know I can't hang off of 40 foot ladders for much longer, and while I'm this engulfed with bees I have so little time to write."[75]

When she gives presentations around the country, Cindy encourages audiences to do swarm removals. It is an easy way to increase one's colonies, she advises. Depending on whether the swarm has made a hive, a beekeeper can take home lots of brood and honey.

When asked if she learned special skills before doing swarm removals, Cindy shakes her head decisively: "If the boys can do it, so can I."

In every swarm removal, Cindy charges a flat rate, does her own repair work, and refuses to use chemicals. She uses a vacuum in which bees get sucked into a secondary source, generally a bucket with comb. One last trick of the trade is that she uses hand saws, such as a sheetrock saw, rather than power saws to cut into sheetrock. Bees cannot groom sheetrock dust off their wings, and power tools blow it all over the place. "It's quicker to use the power saws," she says, "but it's also more dangerous to both the bees and the structure. Power saws on walls are a good way to cut into electrical wiring and plumb-

Cindy Bee, Georgia, United States. Courtesy of Cindy Bee.

ing. I've never done either so I haven't had to learn rewiring. But I have replaced siding, flooring, sheet rock, roofing, shingles, down spouts, soffits, paneling, and other woodwork."

When asked about possible discrimination, Cindy answers forthrightly: "Only a few times did I have people who didn't really trust me once I got to a job site. On all occasions however, once they saw the work I did—the bees that I removed, the repairs I made, and the finished look of their home, everything changed to a much more amiable relationship." She notes: "Women especially voiced their approval that a woman was doing such work and many, many times they mentioned that they were glad they'd hired a woman to do it. Some were incredulous that I worked by myself or that I was willing to climb ladders, use saws, or just deal with the physical work of it all. But without exception the women were always supportive and wonderful to work for."[76]

It is easy to cut corners when doing swarm removal. Who is really going to know, after all, if you haven't used insulation to fill cavities? And will people who have been terrified of having bees inside the house complain about your leaving a little clutter? But Cindy was not interested in the shortcuts that men in construction take all the time: "It really mattered to me that I left a site looking at least as good, and better if possible, than I found it. I don't know how much of this is because I'm a woman and how much of it might be a difference in personality, but I suspect much of it is the gender aspect."

Cindy cuts to the chase: "So many times I've seen men work quick and dirty so they could move on to another job. This is why they get hurt so much. . . . Working with sticky honey on ladder rungs, hands, tools, and buckets is a disaster waiting to happen. Add to that running saws while on ladders amid upset bees and it could get a bit worrisome to think about possible emergency personnel responding."

So I wondered whether Cindy had had difficulties negotiating swarm removals. With her usual candor, she replied: "I suspect I lost bids on jobs when talking to men, and I'm sure there were times when they asked me how I was going to do a removal just so they could do it themselves. For those guys I had no sympathy because I'm sure they got in over their heads. The times that upset me were when I'd actually get out to a job, especially if it was far away, and men would decide they didn't want me to do the work after I'd told them where the hive was located." She learned quickly, though: "That only happened a few times before I made them commit to the job over the phone."

As far as Cindy is concerned, sexism can go both ways. Cindy and a fellow swarm removal specialist named Bill Owens would trade experiences about bidding on a project: "There were times when we both had bid on a job and he won because men told him they didn't want a woman working for them. However, I think I experienced the same with the other side of the issue—women calling us both and choosing me over Bill because of gender. There are probably other variables here too such as how one explains things to the customer, but in the end I really think a lot of it boiled down to gender." Often, she says, women were the ones responsible for keeping up the house: "This seemed to include scheduling bee removal, although I don't think most people realized at first how extensive a project it was going to be."[77]

Cindy has now, to use a Georgia colloquialism, "finally jerked a knot in my own tail" and become a writer. Not surprisingly, her first book, *Honey Bee Removal*, coauthored with Bill Owens and released in 2011, is "a step-by-step guide for anyone who wants to remove honeybees from structures." But a writer is always writing, even while atop a ladder, and Cindy has been recording oral narratives of beekeepers as well as completing a master's degree in creative writing. She sums up her transition: "Doing removals was a wonderful experience. . . . I know once the swarms start happening there'll be this pang in my heart to get out the ole bee vac. In the meantime though, I'm having a great time writing about it, and sometimes that's better than being in the middle of it all anyway."[78]

POLLINATION MANAGEMENT

Lora Morandin provides one of the most exciting new voices in North America. Her research is focused on genetically modified crops. As a graduate student working in Alberta, she focused on the problems an organic Mennonite community that grows organic rapeseed had been having. Genetically modified fields have very few marginal areas, systemic pesticides, and blooms that last only last two to three weeks. Bees need marginal areas to sustain themselves. So Morandin set up a pattern in which marginal areas were included in genetically modified fields. These areas attracted more bees, leading to more seed set. From her calculations on one beekeeper's farm, she was able to determine that his earnings would increase by $13,000 if he included some marginal areas where weeds could grow. This estimate did not address other conditions such as drought or rainfall during pollination periods, but it is a fascinating example of the new information that pollination studies can offer industrial agriculturalists.[79]

KIDS AND BEES, HONEY QUEENS, AND HONEY STANDARDS

Kim Lehman walked away from a profession in library services and during the past four years has traveled around Texas and the United States presenting honey bee and story/music programs to children. She does puppet shows, sings, coordinates activities, and seamlessly melds wildly varied audiences together in her focus on honey bees. One year, Lehman's Vietnamese pot-bellied pig Blossom was photographed for the local paper in her bee suit, as the Texas-Sized Honey Bee. And yes, for the right treat, Blossom could do a waggle dance.

National audiences know Lehman's design of the quirky and colorful children's page in *Bee Culture* magazine. As the creator of the children's club Bee Buddies, Lehman sends members stickers and prizes the old-fashioned way—through the postal service. Bee Buddies has almost eight hundred members from forty-five states and Canada.

Every year, Lehman coordinates the Kids and Bees program for the American Beekeeping Federation (ABF) conference, which also involves the ABF Honey Queen program and public libraries. Lehman does workshops for teenagers on beeswax and speaks to adults on apitherapy—that is, the health benefits of honey bee products, including honey, pollen, propolis, venom, and wax. "I feel I can make a greater impact in bringing joy to children by training teachers and librarians how to use the arts with their students," she states.[80]

Lehman's outreach needs more sponsorship or nonprofit status to expand the scope of children's bee programs and the Bee Buddies club. But she faces a challenge in how people are socialized. In the past 150 years, so many women have become teachers that people expect women to provide education. "Education is a female dominated career," she says. "It's the opposite for performing. Men tend to be better at the confidence needed to be a performer. It is not uncommon to work beside a man where the man gets paid and people expect the women to do it for free or lesser pay. Men tend to rise up in the ranks very quickly while women flounder more."

And Lehman has faced challenges in how *she* has been socialized. Lehman has been actively involved in the theater and indepen-
dent-film communities for twenty years, so she is no stranger to performance. But learning how to negotiate her career transition from
a service-oriented to a performance-oriented profession has meant
making an internal shift: "I've been able to transfer my sense of injus-
tice by looking at the situation in a more constructive way. I've been
learning from my generous male colleagues. I am just beginning to
really succeed as a performer. It is very exciting. It all stems back to
believing you can do it."

Just one experience as a volunteer at the Kids and Bees program
at the ABF conference in Louisville, Kentucky, in 2005 convinced
me of the need for a national children's honey bee education program. On a cold January morning, the American honey queens and
princesses were at the public library, frantically preparing their fact
sheets, activities, and demonstrations.

The event was scheduled to start at 10:00 a.m., and two hundred
people waited for it to begin. For the two hours of the event, a massive rush of kids gleefully bounced up and down the limestone stair-
cases like pogo sticks, tugging parents and grandparents behind like
hapless balloons caught in the wind. Chubby cheeks were painted
with honey bees and hexagons. Beeswax candles made by impatient
children's hands warped at angles that defied geometric definition.
Honey varietals were sampled. Children eagerly bounced upstairs,
downstairs, to this room, that room, this table, no, that table. By the
end of the program, children wilted into coats and parents' arms, the
last vestiges of adrenaline leaving their systems.

Lehman was coordinating this event, but the major attraction
was the American honey queens manning the tables. The honey
queen program fills a gap between children and the general audience. When the program first began in the 1950s, the goal had been
for the queens to generate honey sales. But current trends started by
Patty Sundberg and Anna Kettlewell have taken the honey queens in
a new direction, highlighting the queens' role as educational representatives of the industry, with less emphasis on sales and marketing.

Contestants for the ABF honey queen program need to know the basics of beekeeping as well as the domestic uses of honey for the many demonstrations they do around the country. During the 1980s Carol Holcombe, Tennessee honey princess, won every smoker-starting contest in the Tennessee region. Although the program has moved away from some of those skill-based activities, culinary skills are emphasized now, perhaps leading to an increase in sales. The honey queens demonstrate pollinated fruit, offer samples of honey varietals, talk about observation hives, and answer questions about honey and allergies. The honey queens and princesses provide an invaluable service for general audiences. Elaine Holcombe, a long-time sponsor and volunteer with the program, states simply: "I can call the radio station and schedule a honey princess in a minute, whereas if I simply call and say that the beekeeper convention is in town, no one shows any interest."[81]

The honey queen program offers participants a chance to travel around the world representing the honey industry. They take part in the Apimondia shows, local fairs, and public-school events. It is not unusual for each participant to talk to thousands of people in the course of her tenure.

These pageants can become family affairs. Jill Clark was honey queen and is now vice president of Gamber Container. Elaine Holcombe had three daughters, Melissa, Carol, and Libby, who each served as either a honey queen or a princess. "We told them early on, if you agree to do this, the honey queen program has to be first priority," she says. "You'll have to be available for conferences, speaking engagements, and travel." For one Holcombe daughter, it made "all the difference in the world" in terms of letting her become comfortable with people and crowds. As the bee world deals with mites, importation scandals, and colony collapse disorder, the honey queens are much-needed faces for the bee industry.

Other faces are calling for national honey standards, filling a gap in our lack of education programs for the general audience. Nancy Gentry and Virginia Webb are providing leadership in this aspect of the bee industry. Along with others, Gentry is leading state-directed legislation for honey standards in Florida. Webb, a two-time World Honey Show champion, lets her honey do the talking in Georgia.

Malcolm Sanford calls Gentry a "human dynamo" in one of his articles about honey adulteration.[82] Sometimes, the most effective dynamo is the patient, polite presence of someone who will not budge. Part of such a force is recognizing the futility of arguing with someone determined to stand in the way. Another part is seeing a window of opportunity and knowing intuitively that it may close if you don't persist. In defining this difficult art of negotiating, Gentry says women have had to learn to be "short on patience and long on perseverance."[83] I think that is a pretty good definition of a dynamo.

As countries have begun creating a standard called "country of origin labeling" (COOL), Gentry represents true leadership in arguing for a domestic honey standard in the United States. This quest has been a difficult dance between scientists, attorneys, and others working with the US Food and Drug Administration (FDA) to approve a basic definition of honey that would meet the approval of the International Codex Alimentarius Commission, created in 1963 by the Food and Agriculture Organization of the United Nations and the World Health Organization.

Since then, the National Honey Board has been spearheading efforts to see a US standard of honey. The major honey groups finally agreed to the US version, the "1987/2001 International Revised Codex Standard of Identity of Honey with Certain Deviations," which they sent to the FDA in March 2006. Everything seemed to be in place. But ironically, the FDA happened to lose a honey adulteration case because there was no acceptable standard of identity.

International Codex Alimentarius Commission

- Defines honey in different floral varieties: multifloral, single floral, and honey dew
- Suggests that sugars cannot be more than 13 percent
- Suggests testing to catch metals and sulfides

Alarmed by the fifty to sixty million pounds of adulterated honey coming into the United States, Gentry began a process in which the state of Florida passed a codex by rulemaking procedures rather than legislation. "I stepped into the picture in August 2007 when I heard about the FDA refusal," she relates. "I knew that a state could pass restrictions, adopt rules/regulations etc. to protect their citizenry without waiting on the federal government. So I proposed that if Florida could get a honey standard, adopted July 14, 2009, then other states would do the same."[84]

Now, "citizens of Florida finally possess the constitutional right to access the civil courts where sellers of adulterated honey can be sued for selling a product which does not conform to the compositional standard for honey as established in the state."[85]

Instead of focusing on a national standard, Gentry has focused her efforts on individual states. Rather than trying to prove criminal intent of honey adulteration, why not provide a way for civil suits to be heard? Then the burden of proof would be less and injured parties would have a greater chance of being compensated. Having a standard also would help in those cases when someone buys a barrel of honey that, when tested, does not meet the codex standard. The seller could then face a civil court and the buyer would be compensated for his or her losses. "Whether you pack one pound or a million, you need a standard in your state," sums up Gentry.

There are some who do not agree. After all, the closing of the bonding loophole through an act of Congress, coupled with the increased antidumping rates on Chinese honey exporters, has virtually eliminated Chinese honey from legally entering the US market. According to Ron Phipps, "The Chinese honey industry's ability to ship to the U.S. has rapidly diminished."[86]

But the problem of illegally imported honey is very real. Richard Pasco of the Honest Honey Initiative states: "Record levels of honey are now being imported into the United States from India, Vietnam, Thailand, Taiwan, Malaysia and Indonesia. These last three countries do not have commercial beekeeping industries, so these countries do not have the capacity to produce and export 35.5 million pounds of honey to the United States in 2008 or any year. . . . Transshipment of Chinese-origin honey through other countries is not the only problem. Chinese shippers and others are also misdescribing honey as blended syrup, honey syrup, and malt sweetener to avoid paying the antidumping duty."[87] Presently, at least thirteen states are considering adopting standards, and more are lining up. "It passed in California. It sits on a governor's desk in Wisconsin," says Gentry, "and I know the push is on in Texas, Utah, Montana, Minnesota, Pennsylvania, New York, Ohio, North Carolina, and Georgia. More states have contacted me, so the number grows. Once we have about three, I think the pressure will be on in every state."

If more states adopt the Revised Codex, Gentry will have led the first comprehensive initiative to standardize the discussion about honey in a century. "Isn't there some way to finish what Charles Dadant started? We are long overdue."

World Honey Show–winner Virginia Webb also would like more consistency among national honey shows. Commenting on United States shows, Webb explained: "There's no standardization for honey shows. The judging is inconsistent. There's no US National Honey Show."[88]

Just one honey-judging class with the charismatic Michael Young, MBE, from Belfast, Ireland, transformed Webb's attitude toward honey. "From this encounter with a true master of honey judging, I wanted to raise the bar of honey show competition here in the US," Webb explains. Webb and her husband, Carl, from the upper Soque River basin in Georgia, traveled to England and Germany to train for honey shows.

The Webbs' hard work paid off. The Fortieth International Apicultural Congress, Apimondia, had its first world honey show in 2005 in Dublin, Ireland. It was the first time that Apimondia had sponsored such a show, and there were four thousand jars entered. "I think the show organizers were a bit surprised at the world interest," Webb said.

Part of that interest may have been how the Ireland show was organized. There were quite a few categories. The judging seemed to be more even and consistent. Each class had three judges: an Irish judge, a European Union judge, and an English judge. But each set judged only one class. All entries had to be tested in France, have security seals, and be mailed to Ireland before the competition. "I think the Apimondia World Honey Show was trying to bring honey to the world platform. It brought a lot of new challenges to Ireland," Webb reflected.

Encouraged by her win at the Ireland Apimondia, Webb bided her time until 2009, the next World Honey Show, in France. The entries for this show took more stamina, perseverance, and patience, as well as more than a few pennies. Even though France had simpler rules, to hear Webb tell about her preparation, there was nothing simple about winning another title.

"There were just three categories: multifloral, single floral, and geographical," she explains. "We entered in six categories, but every jar had to be tested. Since each sample was $90, each disqualification was expensive." Three Webb jars were disqualified. Webb's sourwood honey had higher invertase content than other types of honey, so judges assumed that she had heated it. Since higher levels of invertase occur naturally in some honeys, Webb had to have US scientists write letters of confirmation regarding the chemical composition unique to her sourwood honey, from a native Appalachian tree.

The entries had to be sent in three months before the competition. Webb found herself competing against different properties, such as crystallized honey, creamed honey, cut honey, and so on. Other oddities included requirements for photos of topography and the flowers from which honey was produced; in addition, the jars had to be sent to preselected laboratories that were completely overwhelmed with the honey entries.

In fact, Webb thought her entries were disqualified until the very last day of Apimondia. Then, during the very last event, she found out that her display had won the highest award possible—an overall award. One would think that winning two world honey shows would temper some of Webb's ambitions. But to hear her describe her philosophy, it is clear why she's a winner: "My goal is for every jar to be a prize jar. There is so much funny honey on the market."

Since the European Union banned honey containing chloramphenicol in 2002, US beekeepers have been more proactive in recognizing and discussing the need for a national honey standard, but North America still has much progress to make when it comes to defining national honey standards.

Fortunately, then, First Lady Michelle Obama insisted on having a beehive at the White House in 2009. Obama is certainly not the first female leader to use her position to support beekeeping. Other major female figureheads have publicly espoused beekeeping through the centuries. Catherine the Great of Russia and Queen Elizabeth I of England used their political power to support beekeepers, as has Queen Noor of Jordan, and Maria Theresa started the first bee school in Austria. In the United States, former Texas governor Ann Richards kept bees, and Abigail Adams marveled at the bees' division of labor. But Obama is the first one to make a direct link between national health care, diet, and honey bees. In establishing a platform in which she addresses childhood obesity, Obama begins a complicated and proactive chain of preventive measures, starting with healthy, bee-pollinated fruits and vegetables.[89] Obama was so proud of her hive that White House honey was the signature gift to members' spouses attending the 2009 G8 summit.

COOPERATIVES AND TECHNOLOGY TEAMS

Cooperatives and technology teams bring creative solutions to North American beekeeping. Las Cachanillas in Mexico and the Bee Girls of the Ontario Beekeepers Association in Canada represent new approaches that combine resources, share technology, and change perceptions about how team-oriented beekeeping can work.

The cooperative idea evolved during the twentieth century before federally subsidized industrial agriculture became the conventional form of agriculture. Industrial agriculture is so multilayered that economic swings are less disastrous; there is more of a safety network. Cooperatives, on the other hand, feel the severity of economic swings, so when there is a downswing, "cooperators take it upon themselves to go their own way, to the detriment of the cooperative," explains Malcolm Sanford.[90]

Because cooperatives often are loosely organized, marketing has not been well developed and returns generally have been low. So many co-ops have had few incentives to keep members through the difficult economic times. But a women's cooperative in Mexico has had success. Although honey bees are not native to Mexico, they do very well if given proper maintenance and placed in fields with appropriate irrigation. With the increasing regulations in the United States, Mexico prefers to export honey to Europe, which until 2008 had a strong currency and consistent demand for honey. In fact, Mexico exports more honey than it eats.

Conversely, Mexico often imports beeswax, neglecting to capitalize on its beeswax production as much as it could. Las Cachanillas, a cooperative for women in Mexicali, hopes to change these trends so that more people eat honey and support their local cooperative.

Las Cachanillas. In spring 2006, ocotillo cacti stalks snake their way out of the rocky Mexican ground, defiantly offering to the sun their brilliant red flowers. They will not be overshadowed by the barren landscape. In much the same way, Las Cachanillas operates as a co-op, a group of six or seven women beekeepers in a country where 97 percent of the beekeepers are male and located in the southern climes where there is more rain and foliage.

Nothing comes easy—that much is clear to me as I get out the car in Mexicali. Not the water. Not the bees. Not the money. Instead, everything is as if it is scraped by fingernails from the dry, sandy ground, and then held only briefly.

The cachanilla plant is one of those diverse desert plants easily overlooked in this environment. A silvery-gray member of the sage family, its lavender flower attracts bees. It grows only in this difficult

Las Cachanillas, Mexico. Tammy Horn visiting the Las Cachanillas cooperative in Mexicali, Mexico: Josephina, Rebecca, Chata, and Tita. Courtesy of
Elizabeth Burpee.

terrain, and I train my eyes to look on it as if it can somehow numb
the poverty I see in Mexicali. It is this flower that a group of women
have taken as their logo for their cooperative.

Each woman blooms a little when asked what her best experiences have been since becoming a beekeeper. Felicitas Castaneda
Ruano, who prefers to be called Chata, says that when she first joined,
her husband discouraged her. "You are not well," he said. "You
should be at home." But she responded by pointing out her service in
the church. Now, at fifty-eight, she has become an example to other
people in the community that one can start new enterprises. Chata
has since started her own line of lotions and creams in addition to the
regular responsibilities of maintaining her hives, which are painted a
pale pink. Her husband encourages her to work more. He even brings
her coffee in bed occasionally.

Tita's softly measured words say more than the others. She
speaks not of a supportive spouse but of her children, her younger

sons especially. "I never dreamed that I could get an income this way," she says. She's thirty-seven and has four sons. Her two younger sons, Lister and Eduardo, who both love animals, help her because she is allergic to bee stings.

Rebecca Martinez Rodriguez's husband initially tried to prevent her from joining the co-op. Now, when she goes out, he often goes with her. When swarms are in trees, the neighborhood kids come and get him. He has designed a cart for her to haul beekeeping supplies and hives because "the whole world comes to tell him that his wife is a beekeeper," she says. Her hives are painted peach.

Because of her work in the co-op, Rebecca has traveled to Wisconsin and around Mexico. She loves having some income and working with kids. Her daughter is seventeen and helps every now and then.

Josephina Valadez is the most talkative woman of the group. "It is important that our community see us doing this," she begins. "All of us have more pride than we did in the past. When you start a project, you start with nothing, so you don't know the potential you have. But now, the government pays attention to us. It invites us to meetings."

The effect of this federal support on the group is noticeable. "One thing leads to another," Josephina explains. The key to the project is self-esteem. "We no longer wear our hair in little gray braids or wear little blankets. Now we color our hair and we go into the community." Her pride has affected her marriage. Initially, her husband liked to have "his queen in the house, his honey in the hive," she says. But after Josephina became active, "he became . . ." She pauses, searching for words.

"Like peacocks!" crows Gabriel, the extension specialist, from the back room, nailing together frames.

"It's true," Josephina confirms. "He was proud of me in a way he hadn't been before." Furthermore, Josephina's kids are the best promoters. She could tell them, "Here in the house, I'm just your mother, but outside, I'm somebody important." At forty-eight, her eldest son helps her, and her grandson will help her make cookies. Her hives are green. Beekeeping has brought them closer together.

Rebecca plans to finish primary school. At forty-seven, she's learned that "school is really a lot of things you know: it's just arranged differently."

On the visit to the hives, we happen upon some teenagers out hunting rattlesnakes for sport. One teenager is holding a four-foot rattlesnake, still alive, writhing from his grasp. Two mutilated pieces of meat are beyond recognition. A long look passes between Gabriel and the teen holding the rattlesnake. In the distance sit the pink, lavender, green, yellow, and blue hives. The teen finally drops the snake, and the car peels out. The silence settles like particulate matter. Finally, once the dust clears, we can see the hives again. "There is order," Josephina says with finality. "These teens need to see that there is something other than killing for fun. That's why I do this. When we get up in the morning, we say we are going to therapy because that is what beekeeping is like for us."

Josephina is the one who challenges me the most. She wants suggestions for better beekeeping, tips, anything I can offer. I try to accommodate. Before I know it, we are trading ideas for varroa controls. In describing the complexities of the bottom board, I suddenly realize that these women will not blithely order one as I have done from a local bee supply company, but will try to make it.

The Bee Girls of the Ontario Beekeepers Association. At the opposite extreme in terms of technology and climate, the Canadian Bee Girls—Janet Tam, Melanie Kempers, and Alison Van Allen of the Ontario Beekeepers Association—form a technology team. Their responsibilities encompass four areas: honey bee disease control, education, queen breeding selection, and integrated pest management. A mite scouting program was initiated in which a beekeeper "pays the Bee Girls to take samples of bees from various hives in order to test for various diseases including American foulbrood, European foulbrood and nosema. The individual hive costs $2.00 for Varroa mites, $2.50 for tracheal mites, and $8.00 for Nosema."[91]

Although the need for such immediate technological assistance is clear, the Bee Girls' funding comes from three lines: the association, grants, and private donations. The fact that the association has spearheaded this new type of group is important, because it signals a shift in how beekeepers must define their approach to challenges. The association works with federal and state entities but also provides funds for its own members. In fact, the tech transfer team was established in the early 1990s with the support of Dr. Medhat Nasr.[92]

Having an all-female tech transfer team certainly gets attention "but does not necessarily make it easier to communicate with the beekeepers," explains Janet Tam. "It really depends on the people on the team. We have no problem with having a male team member; it has just always worked out this way, so far." Because they are contract employees, there is no maternity leave: "If we leave, we can be replaced, either temporarily or permanently."[93]

"Food safety is one of our main concerns," Tam continues. "We do not recommend any beekeeping management which would compromise food safety, and we make sure to educate beekeepers, new beekeepers and the general public regarding this issue."

According to Tam, "Most people, not just women, don't realize that jobs like ours exist. Anyone interested in a job like this should know one of our favorite sayings: 'Expect the Unexpected.' There are always new developments in the industry, emergency issues to react to, and new tasks the industry wants completed."

However, the same versatility and unpredictability are what she likes about the job: "Sometimes the surprises are unpleasant, but most of the resulting tasks we are familiar with. What we are doing or working on changes throughout the year, from computer/desk work with data/reports/proposals, to lab tests and analyzing/dissecting/counting to fieldwork in the bee yards and travelling around the province to different bee operations to speaking at local/provincial/national meetings and teaching workshops."

The democratic principles that define Canada and the United States have meant that the agriculture and apiculture are starting to be led by women. In part, the rules by which women play are much more transparent than in other countries because of affirmative action laws. New cooperative models can help level the playing field by making rules more consistent, too. But the real key is that North American women beekeepers have a long history of creativity and innovation. As the challenges facing beekeepers continue to increase, creativity will remain this continent's strong point in providing for pollinators.

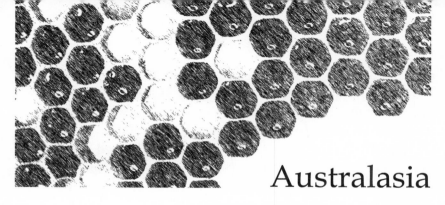

Australasia

A Cornelian Continent

Being a migratory apiarist is no game for a woman.
—Kylie Tennant, *The Honey Flow,* 1956

Honey bees are not native to Australia, and from its inception as a penal colony in the late seventeenth century, the English arrivals to the country were never charged with that Christian ideal of creating a land of milk and honey. There were no ancient iconic goddesses as with Asia or cave drawings of honey hunts as with Europe, India, or Africa. Yet honey bees did well in Australia if they were able to survive the difficult oceanic journey. Because England imposed a powerful cultural template on Australia and New Zealand, Australian women beekeepers had conventional feminine paradigms within which to work, such as the good wife, the angel of the house, and the new woman. But as in North America, frontier conditions never met British conventions for women. Australia also had influxes of international immigrants. Major waves of immigrants from Europe in the post–World War II period, Vietnam in the 1970s, and India in the 1990s meant that Australia's codes for women have never been static.

The most articulate nineteenth-century writer about bee transportation difficulties was Mary Bussell, who emigrated from England to Western Australia in 1834. The Swan River Colony, now the thriving city of Perth, began in 1830 and had all the attendant difficulties of pioneer life. Bussell wrote on March 3, 1834, "I am very anxious about my bees. So many have died within the last day or two."[1]

On March 4, 1834, Bussell wrote, "I have cleared away all my poor dead bees. From the number, I believe very few more could be in the hive and I reproach myself for bringing them away, but to die at sea. I have been obliged to remove the bees that are dead from the hive once or twice, since on one occasion a great many of the poor little things revived and at night returned to our scuttle—bees in every direction."

These bees were also cranky: "Mama, Mr. Sherratt's children, and myself were dreadfully stung, nor did we succeed in saving any of them. The few we caught died before daylight when I got up to return them to the hive."

The bees' days were numbered. When Bussell arrived at the Swan River Colony, she wrote that "Foot's dog and cat" were the "only live things to be landed."

Five years later, Mary Bumby sailed for New Zealand with her brother John Bumby, who was to become the new superintendent of missions in that country. Continuing a long tradition of bees, migration, and Christianity, Bumby intended to serve as her brother's housekeeper. The sailing ship *James* left Gravesend for New Zealand on September 20, 1838, and reached the Hokianga River in March 1839. Miss Bumby's hives most probably survived the journey. Although her diary offers graphic details about her introduction to a new culture, it is far too short for an accurate reflection of beekeeping.

During the same time, Lady Eliza Elliot Hobson, wife of Governor William Hobson, arrived in New South Wales, Australia, in 1840 with bee hives. According to Peter Barrett, "The Hobsons had brought bees with them from Sydney in 1840, though Rev. Taylor states that this hive did not increase." No records of Lady Hobson's beekeeping experiences exist, however. She and her children, one son and four daughters, returned to England in 1843.[2]

In 1842 Elizabeth Macarthur kept bees on her verandah while she increased merino sheep herds. Her papers contain descriptions of top-bar box hives. One of the more famous English beekeeper-missionaries, William Cotton, "paid a visit to Mrs. Macarthur at Parramatta who has a capital apiary." He ends his diary entry noting a request "begging [Mrs. Macarthur] to fulfill her promise of sending me some Bees."[3] We do not know whether she followed up with her promise, although Cotton returned to England.

Mary Ann Allom received the most recognition for her beekeeping, specifically in her successes in transporting bees from England to New Zealand. The Royal Society of Arts and Commerce in 1845 awarded her its silver Isis medal. Fortunately, her daughter, Amy Storr, recorded the preparations that Allom made for transportation. She wrote, "I well remember the months of anxious planning and experimenting with bees on top of our house in Bloomsbury, carried on by her before she perfected her scheme for their safe transit. The special hive containing the bees was made under her instructions by Messrs. Neighbour and Sons of Holburn."[4]

The minutes from the Royal Society of Arts and Commerce meeting read as follows: "After passing the Bay of Biscay, the bees were fed with 2/3 honey and 1/3 water. The whole arrived safely in the colony, and wax, the first produce of Bees in New Zealand has been presented by Mrs. Allom to the Society. . . . The Chairman considered that Mrs. Allom's example would in all probability lead to the introduction of Bees into other parts of the world where they might be of great service." As proof that her bees had survived, Allom offered "the first piece of beeswax made by these bees . . . to the Royal Society of Arts, with a small model showing the way in which they had traveled."[5]

In the ensuing years, women kept honey bees with very little fanfare. There are few records, but one key indicator is that the Apiary Society of Victoria, Melbourne, adopted rules and bylaws on February 1, 1861. Among the rules, it stated that "the Society shall be constituted of Ordinary, Honorary, and Corresponding Members; and Ladies shall be eligible for membership.[6] Who these ladies were, we have no idea. No records exist to suggest anyone applied for membership.

Only later in the twentieth century do we find more records of women beekeepers, after women's suffrage had been passed in Australia and New Zealand. New Zealand offered women the right to vote in 1893. The Australian Women's Suffrage Society formed in 1889. Women were granted the right to vote in federal elections in 1902, after the implementation of the Commonwealth Franchise Act. However, women did not have the right to vote in all state elections until 1910, when Victoria was the last state to decide to grant women voting rights. This state was also the last to vote women the right to stand for parliament, in 1923.

The Australasian beekeeping environment is easier to document than women's beekeeping participation. Rainfall has been and continues to be the most important factor in Australia. It is no understatement when a contemporary government publication, *Commercial Beekeeping in Australia* (2007), notes, "Rainfall is all-important and dictates management practices."[7] In "Letter from Australia," written in 1967, Eva Crane explained that two interrelated factors determine nectar availability: bee forage material and climate, primarily rainfall. "Water must be taken if there is no local source or the bees will die. I have heard of water being hauled 200 miles to colonies in dry country—and of colonies being taken 800 miles to a good pollen source."[8]

Primary nectar sources, such as eucalyptus trees, are generally difficult to judge or predict based on bloom. Compared to conventional flowering trees, eucalyptus trees "tend to carry their flowers on the crown, 20–50 meters above the ground." Such botanical markers make it difficult to tell the quality of the nectar source. Eucalyptus flower buds form one month to two years before the flowers open. Crane explains, "Knowledge of the period for which a species will carry its buds tells [beekeepers] when during the next two years flowering *may* start. But drought can cause the buds to drop before flowering, or a heavy rain or dust storm may knock the buds off. In either circumstance, the expected nectar will not occur."[9] Crane's opinion combines the mind of a scientist and the analytic skills of a beekeeper.

Hear instead Australian native Kylie Tennant describe the eucalyptus as the "crowning glory of this country, and their blossoming, when the great furls of tree-tops look as though they have been showered with snow, their clean scent and color, the way they reflect the hard, hot light, their careless, lopsided shaggy ugliness, is more than any graceful precision of gardening. They are the creatures of the wilderness, and their flowers . . . are a foam on the tree's rejoicing."[10]

Even when eucalyptus trees bloom and yield nectar, sudden weather changes will cause the trees to cut off the nectar flow. Another complication is the nature of eucalyptus pollen. Some eucalyptus pollen is collected by honey bees and used for rearing brood, but other pollen is collected and not used by the bees. So other pollen sources need to be immediately present to offset the nectar flows. "It is the pollen supply that limits the use of the flowers by beekeepers," Crane explains.

But eucalyptus trees have value as a honey plant, as Tennant writes in *The Honey Flow:* "Eucalypts are scraggy raggy trees, writhing plumy trees with projecting skeleton boughs and elbows, broken jagged spars and left-over sticks like old hairpins and hatpins in their feathers, great limbs wrenched off, untidy scattered timber, leathery old leaves and tender green-gold tips. They are take-it-or-leave-it trees with their pants and braces hanging down and their white arms emerging from dirty old grey flannel sweat-shirts of bark."[11] Kylie Tennant would know. When the Great Depression struck Australia in the 1930s, she became a migratory beekeeper. Despite a middle-class upbringing and a private-school education, she chose to document the plight of the poor. She walked six hundred miles from Sydney to Coonabaraban, meeting her future husband, Roddy, who to his credit, nurtured Kylie's writing and political leanings.

Immersing herself in the world of the working poor, Tennant followed itinerant beekeepers, eventually writing *The Honey Flow* in 1956, told from the point of view of a fictional woman named Mallee Herrick. Mallee is named after a honey tree. The first paragraph of *The Honey Flow* is the best description I've heard of migratory beekeeping:

Every time my memory opens its mouth it dribbles roads. Not so much the great bitumen and concrete flanks that cut the mountain spurs and plunge over the edge of the plateaus, but bush tracks that suit a kangaroo or a rogue bullock, but look incredible to drivers who have never had to force a great truck loaded with bee boxes or honey tins through the forests, over corduroys where the forestry gangs have thrown down a few trees to make a footing in a swamp, down the steep creek beds, over places with names like Muldoon's Mistake or The Downfall. Places where we had to make the roads ourselves, taking trucks through the scrub, leaving tyre tracks to be followed later, twisting between trees, places where we were bogged, and the legend was that you jumped as the truck sank unless you wanted to go under the red sand with it, unloading, jacking up, inching her out, loading again, roaring on.[12]

Herrick is in her twenties, determined to make a success of beekeeping. She tries to purchase bees from an elder patriarch, who has decided he wants out of the business. The elder insists that "by the time I pay petrol and running expenses and taxation and the mortgage, I'm still being played for a sucker by a mob of insects."[13]

Inevitably, Herrick learns why the patriarch thinks beekeeping is "no game for a woman." In the novel, Herrick stays for a year with the migratory bee family before she finally decides to part ways with them. The novel ends with ambiguity: Will Herrick get married? Will she run her own bees? We only know that she stays with an older bee friend, determined to put the migratory industry right. "There isn't a worse organized business in the world than the migratory apiarist's. I've learnt now. I know the groundwork. What it needs is the proper footing. Not just this aimless careering round the country. . . . I'm going to have the biggest bee business in this continent. A decent costing system—a central exchange where information can be gathered—distributing centers—men posted at strategic points to take delivery and handle the hives when they arrive . . ."[14] And on this note, the novel ends with Herrick pondering her next step in the outback.

Some of Herrick's frustration with being a woman in a man's world was still prevalent in the 1960s. According to Crane, "Even

Kylie Tennant, Australia. Courtesy of Fairfax Photography.

in 1967 there was still a strong differentiation between the status of men and women, and Australia was the first country I visited where I frequently encountered a gender problem." She continued to explain: "The commercial large-scale beekeepers whom I had been invited to help were all men, who needed much physical strength to handle heavy hives of bees. These beekeepers formed all or most of my audience at Beekeepers' Association meetings, and with them I became a sort of honorary man. With any woman I met I reverted to my biological gender and entered into their lives. At any social gathering, however, the men and women formed two separate groups at a distance from each other and I did not know where I belonged." It should be noted that Australian hives are eight-frame hives but as heavy as ten-frame hives in the United States. Although heavy hives have certainly proved to be a challenge throughout the centuries, Crane ventured to suggest, "The solutions to many of these [problems] seemed to me to depend less on scientific research than on changing some of the ideas and habits of beekeepers."[15]

Aboriginal cultural adjustments to honey bees were noted by writer Annie Elizabeth Wells, who published *Men of the Honey Bee* in 1971. This autobiographical book reflects Wells's experiences as a teacher at a reservation school. In trying to understand how three men could love the same woman in three different ways, Wells had to understand the complicated tribal genealogies based on polygyny, in which men could have several wives at one time.

Wells eventually learned how to tell her tale about a woman named Lindirij: "I followed the thread of a story that grew on me as it was unfolded. Sometimes that thread had to be untangled from a veritable skein of threads, but at other times it lay clear as a rope from boat to anchor." In the prologue, she describes Lindirij's magical effect on men: "To one, she symbolized the welcome light of a camp fire; to another, the beckoning beauty of the evening star; and to the third, the life-giving radiance of Walu, the sun."

Aboriginal Customs of Polygyny

The system of polygyny in harsh landscape meant that kinships were respected across vast distances and groups would be welcomed into another territory if distress forced such assistance. As anthropologist Stephen Mithen explains in *After the Ice:* "A 'cross-cousin' marriage system [is one] in which a man was expected to marry a woman who was his mother's mother's brother's daughter's daughter. As these relations were understandably not easy to find, men often sought partners among groups living hundreds of kilometers away. And as a man could take several wives, he often became related to different families living throughout thousands of square kilometers of desert. Consequently, there was always the possibility of finding kin, and hence access to water and foraging opportunities, in times of need" (324).

Men of the Honey Bee centers around three aboriginal brothers, Maingala, Darangui, and Ngir. In order to gather honey in the Northern Territory, or Arnhem Land, the Aboriginal tribes make pandanus bags. Darangui is the most skilled artist of the three brothers:

> It was a small bag, shaped like a pocket and meant to be worn hanging from its owner's shoulder under his arm-pit, or from his neck, against his chest. This pattern made by feathers, possum-fur string, and colored pigments—woven or painted on the outside of the bag—showed to those who knew such things not only the wearer's moiety and totem affiliations, but also his country and his kinship relations. . . . The work of Darangui was famous throughout all the area; he could weave a pandanus bag of such smooth perfection that the vinegary native honey could be carried in it without leakage.[16]

Women also made pandanus bags. Wells described the process as a "pleasant task, but one that took time. With expert movements of thumb nails and finger tips the leaves were stripped of prickles along their edges, then peeled apart into papery layers. Each long layer quickly curled inwards and became a delicate straw color, forming a pliable strand ready for twining into light basketry."[17]

Through the course of her childhood in the Northern Territory, Lindirij is promised to Maingala, who takes two wives before Lindirij is old enough to marry. After a lifetime of being a third wife and bearing children, Lindirij takes care of Maingala when he dies.

As a widow, Lindirij faces the possibility of remarriage, having to make a choice between Ngir, who is the older brother, and Darangui, the younger one. In a meeting with the community, Ngir forces his hand, but Darangui has a chance to speak. Reflecting his awareness of the new rights for women, he says:

> I speak another word. This woman Lindirij is mine by our law, for she was wife to my only full-brother. But hear me further. . . . There is coming a new law in this country, and in this new law a woman has the right to her own decision in a matter such as this. I say now, let the woman decide for herself between us—or, if it is her desire to live

without a man let her say so, now. Such is her right in the new law, and what she decides, she should be free to follow.[18]

Lindirij chooses Darangui for his wisdom and because, as she tells Ngir, "Your name is not written in my heart."

Honey gathering of any type was difficult in 2007, the year I went to Australia. A severe drought was in its twelfth year. In addition to the drought, cold weather caused the bees to stay inside. The yellow box budded early, flowering in May instead of October. "It was too cold for bees to fly," stated Bill Weiss, the president of the New South Wales Apiarists' Association.[19] West Australia also suffered the same climactic conditions. The end result was that conventional beekeepers started going back to other careers precisely when Australians developed a world market for their honeys. There were exceptions, including Betty McAdam and her husband, Jim.

Located on Kangaroo Island in southern Australia below Adelaide, Betty and Jim wanted a change in lifestyle; they started with two hives and now manage two hundred. A former office manager for a firm of patent attorneys, Betty writes, "I consider myself atypical of women in beekeeping in Australia in that I was the instigator rather than the supportive partner. My husband, Jim, freely admits that all this pain was certainly not his idea."[20]

There was a learning curve to be negotiated between the two: "I did not anticipate that Jim would become interested in beekeeping from a point of view of problem solving and would offer advice, whether requested to or not. We solved that problem by getting him his own protective gear and we now work together."

"There are divisions of labour," she explains. "I do the swarm catching, public relations, record keeping and queen bee breeding and Jim organizes the equipment, computer programming, maintenance and forward planning. However we work together in the apiaries, extracting and packing honey, because we enjoy doing so."

In response to questions about challenges such as flooding, droughts, and fires, McAdam answers: "We do not actually feel chal-

Betty McAdam, Kangaroo Island, Australia. Courtesy Betty McAdam.

lenged. There are day to day problems to resolve. Seasonal variations mean that honey production fluctuates, but as we have no debt to service, we can live within our means. Events such as becoming bogged while moving hives and having to request assistance to tow us free, or having hives collapse and release hordes of aggravated bees are simply part of life."

She does anticipate one problem for the future, the physical labor involved in working the hives: "We have no difficulty selling all the honey we produce and could expand up to 400 or 500 hives very easily but cannot manage this number effectively without mechanisation, which is not the way we want to work. We are employing a local resident for a day a week, training him to harvest honey or carry out specific seasonal tasks, whilst we oversee operations and concentrate on improving the genetic lines and monitoring hive changes."

Describing the breakdown of various tasks, Betty explains: "As is pretty universal, I get to wash the bee clothes, do the floor cleaning and washing of buckets. I also maintain the records and enter them on the computer. I pay the bills and keep the tax records. Jim designed the system for our use. I am the public face of the business for talking to customers, promotions and sales, television appearances, farmers markets and telephone enquiries." Betty is also the swarm collector, "as this involves traveling to resident's homes or farms and keeping them informed on bee behavior as well as putting on a show and providing photo opportunities."

In addition, Betty says: "I do the dirty, grubby jobs of frame cleaning, sorting and discarding frames, processing equipment through the freezer and cleaning up spider webs. I am also the 'gofer' while we are working hives together, carrying the frames and brushing bees. I manage the uncapping for extractions while Jim attends to the extractor and pump. I do not handle the mechanical equipment or put bee boxes together."

Betty also keeps up with industry goings-on: "I read the reports and regulations issued by industry associations and government associations so as to be aware of any changes that affect beekeeping. I am the contact person for the auditor or any inspector. I am also the Secretary/Treasurer of the Kangaroo Island Beekeepers Association, which involves the keeping of minutes, Association correspondence and a regular Newsletter." Finally, she says: "I breed queen bees for replacement of unsatisfactory genetic lines. I process beeswax into candles and beeswax blocks for sale. I harvest the propolis from collection mats and handle correspondence for sale of raw propolis."

With regard to gender differences in beekeeping, Betty says: "Small scale beekeeping is as easy, or easier, for women to enter than men. Hive management, production and marketing of honey and bee products can all be run from a home base. Men face the same problem as women in scaling up operations to include hive migration and handling heavy supers. There are mechanical aids but you need to increase hive numbers to make such things economic." She expounds: "The fascination of beekeeping is experienced by certain personality types—usually 'hands-on tinkerers' who don't mind getting down and dirty. Making a commercial success of beekeeping depends on organizational talents and hard work."

One thing that Betty believes would be helpful to beekeeping is if governments would "remove obstacles to individuals taking their first steps." According to Betty, "Beekeeping is one primary industry where small scale entry is advisable. Regulations by local government and State governments to prohibit keeping of bees due to perceived public risk need to be weighed against the benefit of bees to pollination and self sufficiency. State government funded training for beekeepers and enforcement of a code of practice for beekeeping to minimize nuisance would be more constructive than prohibition which leads to lack of regulation and education." She notes, "There are instances of objections being lodged to the keeping of bees even in semi rural areas by people a mile or more away claiming risk of danger from foraging bees."

McAdams concludes with a bit of advice that we can all use: "Beekeepers must recognize they are all on the same side. There may be many opinions and methods of working that are valid for the individual. This must not detract from recognition of the common interest."

McAdams's points were reinforced by the women beekeepers I met from New Zealand and Tasmania. If eucalyptus trees are the staple for Australian beekeepers, Tasmania's leatherwood trees make or break beekeepers there. Cradle Mountain Tasmanian Honey won five points out of five in a honey tasting by the *Sydney Morning Herald,* the only honey to place that highly.[21] The Cradle Mountain beekeeper Lindsay Bourke and his wife, Yeonsoon, exhibited their honey in Melbourne, Australia, at Apimondia 2007.

Yeonsoon's story begins not in Tasmania but in Korea, where they met: "Lindsay became a beekeeper when he was 23. He used to buy and rent out a house each year with honey crop proceeds. I met Lindsay 28 years ago in Korea. Lindsay was visiting Korea to market his honey. I was working for the Australian Embassy in Seoul where we met. Lindsay was not successful at exporting honey to Korea that time. I moved to Australia and we were married the following year. We became full time property developers soon after. For the following 20 years, we were property developers as well as hoteliers and restaurateurs."[22] Lindsay reentered the beekeeping industry in 2002 and proved he still had all his skills by producing the best leatherwood honey, awarded by the Tasmanian Beekeepers Association, five times in six years.

But Lindsay's success has a price. For Yeonsoon, that price is being away from home for long periods of time: "The biggest challenge for me is balancing home and business. My younger son is 15 and still at school. I have to be away from home with Lindsay often and I worry about [my son] when I am away."

In addition, Yeonsoon explains, "I am not a hands-on beekeeper but I will help Lindsay if he needs it. I upgraded my licence to drive a medium rigged truck. I drove the truck in the past to help Lindsay when we shifted the bees so that Lindsay could have a little rest during the long hours of driving. I understand the process of beekeeping and production." Most of the time, however, "I am at my desk or marketing. I manage the administrative, financial and sales areas of the business. These areas are quite exhaustive and include many local, national and international trips each year. We strongly believe in personally representing the company and our products so that clients have a face to the product. Being Korean, my language skills in Korean and Japanese have proven valuable when breaking into these markets. The branding, labelling and presentation of the products take a considerable amount of time and need to be regularly reviewed." During busy periods, though, Yeonsoon has a more hands-on role with the honey packing, which she finds "relaxing and therapeutic."

Yet, for all the worry and stress of arranging logistics, Yeonsoon likes attending conferences and learning new ways of business: "I try to diversify our business and widen the product ranges. We produce not just honey, but organic honey, honey mead, honey ales and honey nectar, honey cordial concentrate. This is to encourage more honey consumption and quicker turnaround time for stock." The Bourkes are now the largest honey producers in Tasmania.

Maureen Maxwell, third-generation Kiwi, agreed to meet me at Apimondia in Melbourne. Maxwell has been described as "a tall woman with a reddish-brown bob that swings in time with her athletic gait," but that leaves out a magnificent, hearty laugh and her gracious ability to tell stories.[23] Raised in the country, she studied industrial architecture at the University of Auckland before going to Europe to be a chef. Returning to New Zealand in 1980, she settled in Waimauku on an old farm with fruit trees. She loved the taste of fresh honey so much that she bought her first hive from a newspaper ad. Within a month, she had a full super. Within six months, she had 273 hives.

Having been a chef at four restaurants, Maxwell opened BeesOnline in 1999 in Waimauku and deliberately kept it casual. "Despite its name, BeesOnline is not an Internet apiary matchmaking service," reassures writer Caroline Campion, "it's a restaurant, shop, and honey production center."[24] The restaurant does not accept reservations so local people can come. The goal is for BeesOnline to offer people a regional experience and education about local food and bees. It is a process she calls "bee theater." "Anyone can walk in free of charge and talk bees," she emphasizes with a nod. BeesOnline received the 2007 Trailblazer Award for architecture and certified organic factory.

Maxwell outlined the challenges facing her when she first began: "No one had been a face for the industry. What became obvious was that we needed a profile. We needed to explain to the world [the importance of organic honey]—we need to pull together as an industry."

Working in conjunction with the environmental awareness movement, in which items that are eco-friendly are identified as green, Maxwell decided to market honey as a green product. She leaned into the table, counting her fingers with the following ways that honey is "triple green": "It looks good, it tastes good, and it is good for the environment." She realized she needed to market honey like she used to market wine. This means educating buyers about

Maureen Maxwell, New Zealand. Courtesy of Maureen Maxwell.

the product through educational labels: single-flower honeys; harvest once a year; working with the weather; terrain makes a difference for something like manuka honey. For example, "Dark honey is like red wine—mineral, herbaceous, earthy, blend of masculine flavors," she explains.

The marketing strategy of BeesOnline represents a shift for New Zealand, a country of about 4.1 million people. Wages are low, and when it comes to education, "People are going off-shore, although the country used to have a solid educational platform." The main income earners have been meat, wool, and dairy products. Although tourism and now honey bees, especially manuka honey, are bringing more money, Maxwell sums up a problem with New Zealand: "We spend our entire lives looking outside."

When asked about her experience as a woman beekeeper in New Zealand, she said that initially, the club members "were not open to sharing knowledge." Still, she and her friend Rae had fun. Since most beekeepers have beards, she and Rae would wear false beards.

The challenges facing New Zealand beekeepers are like those everywhere now. Varroa mites arrived in 2000. Maxwell lost a lot of hives. If American foulbrood surfaces, the policy in New Zealand is to burn hives. There is no government compensation.

Another New Zealand beekeeper, Jane Lorimer, describes her entry into the industry as a slow progression, especially compared to Maxwell. She took four years to decide to get involved. In 1985 Jane had been working on a research degree in biology. She married Tony, whose father needed assistance with construction on his new honey house.

But, she says, "I finally couldn't help myself," with an air of an intelligent person who has been analyzing a game from the sidelines and finally decides to be a game changer. Her interest in biology gave her a curiosity about bees. Initially, she said, "being a woman hurt because there were so few actual beekeepers. Now there are a lot more."

At the time I interviewed Jane, we had just been to see sheep demonstrations at Apimondia in 2007. We were surrounded by tables of different Australian honeys and plenty of female beekeepers from around the world. Jane had just stepped down from being executive president of her association, the second female in their history. "They recognized I had brains," she says simply. "I brought a perspective about value-added products, being proactive rather than reactive. The arrival of varroa mites forced that, funding—all those issues needed leadership." With a thousand colonies, Jane and Tony stay busy, but they are "actively involved with education so that younger people have an image to associate with beekeepers."

Finally, to balance the commercial and entrepreneurial women, Dr. Young Mee Yoon works with the research associated with New Zealand honeys. A graduate in food science from Yonsei University, Michigan State University, and the University of Otago, Young Mee Yoon develops research on honey and its antibacterial properties and regularly attends Bee Products honey standards council meetings. She participated in recent research using honey for medical wound dressings for the company Apimed Medical Honey, which works toward achieving a good management practices GMP system as certified by New Zealand Medicines and Medical Devices Safety Authority. Furthermore, she has been involved with setting up systems and standards for an accredited medical-grade honey supply chain.

Young Mee Yoon writes: "I saw there were big opportunities in research and product development for the New Zealand honey industry. Unlike the meat or dairy industry, it was such an undiscovered and unique New Zealand industry."

When asked about her opinion regarding the challenges facing women, she responded: "I do not think a young woman will have more or different challenges than a young man in this particular industry. It is a well balanced industry for men and women. Quite often I see beekeeping and honey processing are being managed by a husband and wife team; often the wife manages business and general administration while the husband manages the jobs that require more physical strength such as beekeeping with long hours of traveling. Women often have responsibilities looking after their children while their husbands are away."

Hailing from Australia, Grace Pundyk would not qualify as a migratory beekeeper, but she is a migratory honey sampler. In *The Honey Spinner: In Pursuit of Liquid Gold and Vanishing Bees,* Pundyk follows a trail of honeys from Yemen to Turkey to China. She also learns some difficult truths concerning imports and exports, honey contamination, and pesticides. She sums up some of the challenges facing beekeepers: they must address the conundrums created by a combination of ignorance and arrogance that has contributed to our decline in bees.[25]

In her conclusion, Pundyk brilliantly retells the myth of Aristaeus, having him attend the Apimondia congress in Melbourne in 2007. Through this character, Pundyk makes her point that honey is the United Nations of food. It is political, and the problems in the industry are created as much by people as by shifts in bee biology, immunology, and genetics. Australia has positioned itself to export honey bees to the United States. By transshipping package bees, Australia opens rather than closes international questions regarding pest containment. Once the magnitude of the problems concerning honey standards is understood, Aristaeus returns to the realm of divine status. But ordinary mortals do not have this choice. Having wandered the world, Pundyk returned to Australia, but she does not offer easy solutions for the complicated global honey industry.

South America

The Continent of Tomorrow

> To me, beekeeping and queen rearing are both a lifestyle, quite
> different from urban activities where women are more tradition-
> ally involved.
> —Sonia Verettoni, 2008

South America brings this book full circle. South America was
the first continent to undergo a cultural and biological revolution
through the human-assisted migration of African honey bees.
The introduction of the African honey bee, *Apis mellifera scutellata*,
to South America in 1956 has made South America a type of "observation hive" in more ways than one. As the world watched the
northward advance of African honey bees, a generation of women
researchers and beekeepers witnessed two shifts from the grass-
roots level: the honey bee displacing native bees, and the emergence
of women's equity issues.

Although South America was the last continent to address inequities for women, women have been important participants documenting the African honey bee's adaptation. They have been scientists,
Peace Corps instructors, or beekeepers adjusting to opportunities
presented by the African honey bee. Before 1956, South America
was not considered a major center of commercial beekeeping. Today
it is a major exporter of organic honey, a center of research, and an

ever-evolving field laboratory to study genetic lines coexisting with African honey bee genetics. The European Union has set standards for both honey and human rights, and the South American honey "powerhouses" have been reconfiguring their policies in an effort to participate in international communities.

For Brazil, which operated by Spanish-Roman laws and codes, women were considered property until the twentieth century and did not have the right to vote until 1933. The right to divorce was not granted until 1977. Similarly, Argentina, which operated under Napoleonic code until the twentieth century, did not grant women the right to vote until 1947. Women's rights tended to be the last priority, given the volatility of its government. Women could finally divorce in 1986. While these two countries are the emerging honey production centers, other South American countries such as Peru, Bolivia, and Venezuela have been affected by African honey bees and have proved fertile communities for international researchers. These countries have encouraged women to become beekeepers and sponsored international agricultural organizations. Consequently, South American women lead research and queen production and participate in discussions about honey production.

Before I progress much further, I should note that South America had honey bees before 1956. F. Padilla and colleagues speculate that *Apis mellifera mellifera* (German black) and *Apis mellifera iberica* had been brought to Brazil by the end of the eighteenth century. European bees *Apis mellifera ligustica* (Italian) had been taken to South America, presumably by D. Patricio Larrain Gandarrillas in 1848, but they were not proficient honey producers.[1] Consequently, most indigenous communities continued to work with their native stingless bees, *Melipona beechii*. For centuries, indigenous peoples paid their taxes to the Spanish or the Mayan divinities with wax and honey from these ground nesters. "No records remain of any disturbance caused by the introduction of stinging bees to South America when *Apis mellifera* was taken from Europe several centuries ago," Eva Crane, at that time editor of *Bee World*, wrote in 1972. "Their descendants are in general mild-tempered and not always very good honey getters."[2]

But in 1956, African honey bees *Apis mellifera scutellata* arrived, via human-assisted migration, in São Paulo, Brazil. Crane explains: "The unintended release of a few queens of the more aggressive Central Africa bee *A.m. adansonii* [now called *scutellata*] has brought consternation to the whole American continent."[3] Writing with more North American informality, Malcolm Sanford called North and South America the "last Hurrah" for the African honey bee, comparing the insect to human pioneers who crossed prairies and mountains in North America. "From relatively few colonies has emerged a biological revolution of almost mythic proportions," says Sanford, "resulting in an insect migration of thousands of miles in a few short decades, almost saturating tropical America with honey bees."[4]

Despite initial misgivings about African honey bees, according to Sanford, "the African honey bee has now become somewhat of a savior of the beekeeping industry in many countries." Such potential for South America to lead the commercial bee industry had been predicted by Crane in 1972. In her editorial, Crane stated, "Twenty years ago, Latin America was almost unknown in apicultural circles elsewhere. Today it is one of the few regions of the world with a large under-exploited honey producing potential."[5] That potential finally is being taken seriously by a European community wanting organic honey and a North American community that cannot meet its domestic demands.

Brazil

The feminine beekeeping tradition as it was established in Europe and North America—with women keeping bees either in domestic or religious convent environments—did not transfer to Brazil. Women were considered property of their husbands, and their education was not considered an appropriate investment until the nineteenth century. Extending Roman values regarding chastity for women before marriage, Brazil offered few options for women wanting a life without men or children.

High levels of poverty have also persisted through the centuries, and many of those affected are women. This was the case when Crane visited in 1973. Using figures from 1983 in her autobiography *Making a Beeline,* Crane noted that "between 25% and 50% of the [poor] population were illiterate."[6]

Thinking that introducing the more prolific African honey bees would serve as an economic stimulus, Professor Warwick Kerr shipped the bees to São Paulo in 1956. But he was not acting on his own authority, as is so often assumed. Crane explained the logistics of such a study: "[The African honey bee's] introduction was carefully studied by federal and state entomologists there, who were acquainted with the many positive characteristics of the bee. They asked Professor W. E. Kerr, who was going to Angola in order to study stingless bees, under a Rockefeller Travel Grant, to bring back with him about 100 queens." The consensus of this research team was that by breeding selected African bees with European bees, "Brazilian beekeepers could solve low productivity of their colonies and the difficulties associated with introducing queens into nuclei or hives of *mellifera*."[7]

Two shipments of queens arrived, the second with a higher survival rate of queens than the first. According to Crane, "The first shipment of bees arrived from Tanganyika, of which only one queen arrived alive. The second shipment arrived from South Africa. Of 170 queens, forty-nine arrived alive. Twenty-four queens were eliminated because of their defensiveness."[8] In the popular media, these African honey queens were stereotyped with one similarity—defensiveness. In fact, there can be much variation between queens.

One weekend, all the double-screen queen excluders, which prevent queens and drones from leaving the hives, were removed by a local beekeeper unaware of the nature of the bees. The Piracicaba headquarters did not know that the excluders had been removed until ten days later, during which time twenty-six colonies swarmed.

South American beekeepers immediately had to confront the differences between African honey bees and their comparatively gentle stingless bees, the *Melipona scutellaris, Melipona subnitida,* and *Scaptotrigona postica.* Crane documents the effect in Costa Rica in *Making a Beeline:*

> Beekeeping with honey bees had been done [in Costa Rica] until 1983, but the bees then became Africanized and were too difficult to handle. I was very sad to see the disused relics of beekeeping; the hives had stood on two large shaded concrete stands, and stacks of empty hives were piled up everywhere. Nowhere else had I seen evidence of such total collapse of profitable beekeeping as a result of Africanization. In parts of South America, beekeepers had learned how to handle the Africanized bees, and then obtained higher honey yields than they had from European bees which were less well-adapted to the tropics.[9]

As the African honey bees have advanced into North America, fatalities have become fewer because countries have provided better education.

In a side note, Crane seemed to think that far more people faced problems with rabies than with honey bees, but the death rate from rabies was not publicized with the same attention that bees received. Letting numbers make her argument, Crane dryly noted that "every year, 200 people died from rabies in one city alone, which was four times greater than the bee fatality rate, but in my time in Brazil, not one newspaper covered the rabies problem."[10]

Major cultural issues proved obstacles to women becoming beekeepers. In writing about Brazil, Christine Soares summarizes many of the problems plaguing the country in contemporary times: "Bureaucracy, burdensome taxes, and weak enforcement of antitrust and intellectual property laws are blamed for stifling the population's natural entrepreneurial dynamism. A poor school system and high illiteracy rates are the other major barriers to progress most often named."[11]

Nonetheless, some women have triumphed in the apicultural research fields that started under Kerr's leadership. With more than thirty years of experience, Dr. Zilá L. P. Simões, a researcher at the University of São Paulo, Ribeirão Preto, started studying the protein vitellogenin and its effects on the worker bees within a colony. She began as a doctoral student working with Wolf Engels in Tübingen, Germany. A specialist in vitellogenin, Engels had traveled to Warwick Kerr's Genetics Department of the medical school, where Simões was studying as a graduate student. As an undergraduate, Simões had published highly relevant research with an article on the "estimation of the number of sex alleles and queen matings from diploid male frequencies in a population of *Apis mellifera*" in *Genetics*. Consequently, the German government financed Simões's doctoral research. She left Brazil to complete her research in Germany.

Simões had always felt there was something to be understood about vitellogenin's role in the reproductive development and behavioral patterns of worker bees, but she was never able to explain why worker bees would carry vitellogenin when it primarily affects reproductive behavior. Although Simões was not able to unlock the keys to the protein immediately, she cloned and sequenced some of the genetic material for her colleague Norwegian Gro Amdam, who was then finishing her doctorate in experimental biology.

Brazil was the backdrop for Gro Amdam's epiphany when she was studying honey bee nutrition. Before Amdam's research, according to M. E. A. McNeil, it was known that "for a broad range of egg-laying species, vitellogenin is associated with reproduction—as it is for the queen honey bee, who absorbs it through the ovary into egg yolk proteins. But it was a mystery why it was produced at all by non-reproductive workers and why the pollen hoarding workers had more." For Amdam, it didn't make sense: "She thought it could not just be 'evolutionary luggage.'"[12]

When she arrived in Brazil, Amdam's research bees were confiscated. So Amdam reviewed the literature regarding social wasps, a social lineage parallel to that of honey bees. Mary Jane West-Eberhard, an American evolutionary biologist, worked with tropical insects in Costa Rica. Eberhard's research on ovarian development

Zilá Simões in her lab in Brazil. Courtesy of Zilá Simões.

in wasps and its potential effect on social behavior influenced how Amdam looked at the role of vitellogenin in honey bees, specifically the worker bees.

Simões and Amdam worked together in Brazil and Norway to develop a "knock down" method to downregulate the vitellogenin protein in worker bees. The results from their studies indicated that vitellogenin affected hormones, health, and behavior. With Zilá's assistance as an "intellectual parent," Amdam was able to show that reproductive sociality is predictive of social behavior.[13]

Building on the research of Simões and others, Amdam was able to connect vitellogenin with the production of royal jelly by nurse bees, who process it through their hypopharyngeal glands and feed it to brood. As McNeil describes it, "[Gro] further found that vitellogenin suppresses juvenile hormone JH; conversely a drop in vitellogenin results in a rise in juvenile hormone, triggering foraging behavior."[14] Simões and Amdam continue their research and friendship by getting together every other year to discuss knock downs and develop new ideas. In short, they "build understanding together," to use Amdam's words.

Meanwhile, out in the field, Brazilian beekeepers adapted to the African honey bees enough that they could begin to take advantage of their vast acreages and a global demand for honey. With presidential leadership that actively encouraged and promoted science and agriculture, Brazilians have become one of the major players in the bee industry. It is the world's fifth-largest nation in land area and is rich in natural resources.

In 2005 Brazil exported 14,400 tons of honey to the European Union at a value of US$18.9 million.[15] In 2006 the EU suspended the import of Brazilian honey because it expected more quality-control analysis, specifically for antibiotics and heavy metals. But by 2008 Brazil had adopted strict codes for exporting organic honey. Consequently, according to Ron Phipps, "Brazil produces about 40,000 MT [metric tons], half of which is consumed locally and the other half exported. America and Europe will compete on par for Brazilian organic honey. . . . The strong Euro plus the fact that in some European nations 20% of the food is organic has made Europe the preferred destination for Brazilian honey."[16] In short, Phipps summed up, "It is an unprecedented situation."

Women beekeepers will benefit from Brazilian president Lula's

President Lula's Investment in Brazil's Science Cities

President Luiz Inácio Lula da Silva, popularly known as Lula, was behind much of Brazil's cultural change. Determined to stop the expatriation of Brazilian citizens, he actively supported a new type of education, a system of "science cities," in which state-of-the-art campuses are positioned in poor rural areas so that young people can learn and benefit economically. "Science can leverage economic and social transformation," according to Miguel Nicolelis. In an article about Brazil's new direction to a knowledge-based economy, Nicolelis states, "Ninety-nine percent of scientific work doesn't require a Ph.D." (Soares, "Building a Future on Science," 83).

investment in "science cities." Already, major changes in affirmative action are evident. According to Simões, "There are still cultural barriers, though they are normally subtle. Fortunately, when women are evaluated academically by peers, they have the same chance as men in our university system."

Speaking as a laboratory scientist, Simões sums up Brazilian research:

> The most difficult challenge [has been] to obtain consistent funding to allow us to build, equip and run our laboratories. Honey bees are not native to the Americas and how they adapted to and became part of our environment and society is very worthy of study. We also have hundreds of species of native stingless social bees. All of this, together with pioneer work by scientists such as Warwick Kerr, Paulo Nogueira, and Father Jesus Moure, helped form what is now the largest group of bee researchers in the world on one campus. We have 11 professors and several dozen post-docs and graduate students, all working on various aspects of bee research.[17]

When asked about beekeeping opportunities for women, Simões replied: "What ultimately determines whether a woman enters a certain field is job opportunities. If there were more and better opportunities, such as decent-paying public elementary education, more women could become involved."

Argentina: Land of Four-Wheel-Drive Bees

Like Brazil, Argentina is a major participant in world honey exportation. In the 1990s it supplied about 27 percent of total honey trade, approximately one hundred thousand metric tons.[18] At the beginning of this century, Argentina profited from record-high world prices for honey, so much so that during the ten years before 2006 it was "three times the world's largest exporter," according to queen producer Martin Braunstein and his wife, Sonia Verettoni.[19] Braunstein and Verettoni own the Malka Queen Company.

But when it comes to queen production, there is room for improvement on the exporting end. Braunstein and Verettoni exported their first queens to France in 1999. Before that, they state, "the export and import of queens and packages used to be a business more focused between Australia, New Zealand, and Hawaii on the supply side and the European Union, Canada, Middle East, and Far East on the demand side." Nevertheless, they conjecture, "recent findings of the Small Hive Beetles (SHB—*Aethina tumida*) in Australia, along with fears about the possible presence of other pests such as the Asian *Tropilaelaps* mite, have created a niche for an alternative Southern Hemisphere supplier like Argentina."[20]

Having devoted much of their adult lives to developing a queen production business, the husband-and-wife team refer to their Malka queens as "four-wheel-drive bees" because they can make honey anywhere in the world in all kinds of terrain.[21] The couple's success has much to do with the queen breeding techniques they learned in North America.

Verettoni, who has a degree in agronomical engineering, joined Braunstein in his business in 1991. To hear Braunstein tell their story, their courtship seems simple. "I met Sonia at a beekeepers' meeting back in 1991," he says. "I was 23 and she was 24 then. I fall in love with her at first sight." There was the slight inconvenience that Sonia had a fiancé, but eventually, two months later, Martin was her boyfriend.

"From March 1992 until September 1992 we both worked at the Wilbanks Apiaries in Claxton, [Georgia]" he goes on. They married on January 14, 1993, "and worked again one more year for the Wilbanks." While in Claxton from 1992 to 1993, they grafted almost seven hundred queen cell cups a day. Then they came back to Argentina to start their own queen business.[22]

After they returned to Argentina, they imported Italian bees directly from Italy. Sonia Verettoni describes the typical day: "The first years were full of hard work and long hours. Every morning, we would only know when the day started but never when we would be over with our duties. Until 1997 and 1998, respectively, we both had to work as employees in other activities because the income generated

Sonia Verettoni, queen grafting in Argentina. "She is my left and right hand," says her husband, Martin. Courtesy of Martin Braunstein.

by bees was not enough to sustain ourselves. So we devoted the afternoons and the weekends to expand[ing] the number of our beehives. I enjoyed debating with Martin about the best possible methods for queen production; however, my major contribution was problem-solving of technical and mechanical issues."[23]

Martin elaborates the differences: "I think she works harder than me. She takes care of many responsibilities such as handling the breeder queens, the grafting, worker supervision and finally she is a wonderful mother of two kids, Ezequiel and Melissa."[24]

Sonia speaks with refreshing candor about the difficulties of balancing motherhood and queen production: "Being parents was a decision put off for many years into my mid-thirties. I simply could not cope with the demand of the growing business along with the care of a young baby. So, it took nine years since we met each other until my first son Ezequiel was born in 2000. Four years later, Melissa

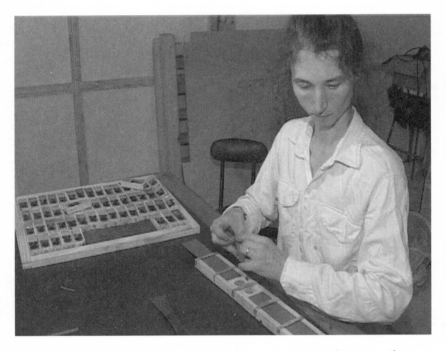

Sonia Verettoni, loading a queen battery box in Argentina. Courtesy of Martin Braunstein.

was born and we completed our family, which also includes two horses, one cat and four dogs."

She continues: "My two children were born in November, which is the peak of spring season in Argentina [the busiest time of queen production season]. I recall working until a week prior to delivery and getting back to work ten days after their birth. I remember an anecdote of October 2, 2000, when about a month prior to the birth of Ezequiel, my husband broke two bones in his right leg at the beginning of bee season. He was fifty-five days semi-immobilized and it was very stressing to perform my duties as well as part of his responsibilities." Their lifestyle had its benefits: "Working and living at our farm and workplace gave me the rare privilege of breast-feeding Ezequiel until he was one year old and Melissa until she was three and a half years old."

Melissa Braunstein, Argentina. Courtesy of Martin Braunstein.

Sonia and Martin have weathered challenges presented by varroa mites—which Braunstein calls "the fourth individual in the colony"—and African honey bees. Because the Malka company is located south of Buenos Aires, Sonia and Martin seem to handle the African honey bees. They write, "African honey bees were found in Argentina back in 1968 in the subtropical province of Misiones next to Brazil. However, during [the past] 38 years their expansion was surprisingly limited only to the subtropical region of Argentina with a clear distinction between a rainy and dry season. All attempts, either purposeful or accidental, to bring [African honey bees] into the temperate humid pampas with four clear defined seasons and with plenty of moisture all year round, were unsuccessful."

In 2006, however, the only true African honey bee populations were "some feral hives located in the rainforests of subtropical northeastern and north-western Argentina." According to Verettoni and Braunstein, "Commercial beekeeping in those regions is negligible, with only a few aboriginal communities keeping a small number of colonies with [African honey bee] traits. Migration of commercial beekeepers—to or from those areas—is very difficult because of the distance from the prairies."[25]

Another major factor affecting Argentine honey production is the use of land for biofuels: "For many years Argentine beekeepers took for granted the widespread availability of prairies. However, the collapse of the cattle industry has put an end to the golden era of beekeeping." Soybean acreage is increasing, and consequently honey producers may have difficulty finding nectar sources. An estimated 17.6 million hectares of soybeans were to be planted in 2008.[26]

Verettoni verifies these statistics: "The expansion of intensive agriculture in special soya beans is deteriorating the prospects of beekeepers in Argentina. Lack of pollen and nectar producing plants are putting an end to expansion of beekeeping in this country."[27]

But women beekeepers, who are still a minority, do not follow the migratory trail. Instead, they "decide to become beekeepers when their kids grow up and leave home or they become involved in scientific research." Sonia remarks: "I enjoy being outdoors in full contact with nature. However, this is of no consolation when I am very stressed during the peak of our queen shipping season!"

Verettoni speculates realistically about her children's possibilities for the future: "My two children will receive the legacy of hard work. I remember my eldest son when he was three years old and he saw with surprise that someone he had just met was not a beekeeper! After seeing his father, mother, and five employees working bees full time, he could not believe that a person in this world was not involved in the bee business. I had to explain to him that some people were not beekeepers."

Peru

Although North American bee researcher Marla Spivak had intended to visit South America only as a happy-go-lucky traveler in the 1970s, she ended up on an exciting but ill-fated jungle trip, sick in a hospital in Quillabamba, Peru. Her doctor, who was a small-scale beekeeper, claimed the peasants on his weekend property were stealing his queen bees. It was an area about fifty miles downriver from the ruins of Machu Picchu and about eleven miles from Quillabamba, Spivak says. The doctor offered Spivak a place to stay if she would solve the mystery of his stolen bees. "I did stay on his property and realized that his bees and all bees in the area were becoming Africanized and beginning to abscond and swarm. No one was stealing queens or bees."[28]

With the mystery solved, why not return to North America? Spivak was destined to go in an unconventional direction. "Eventually I met a family down the road from the doctor's weekend place," she relates. "They were a great family and had lots of land, grew most everything they ate. A farm like this, in this part of Peru, is called a chacra. I loved living there, learning how to harvest, dry and roast coffee beans, how to make chocolate from cacao, how to catch and eat prized guinea pigs."

"I ended up marrying the middle son of that family," Spivak continues. "They spoke no English, had no running water, no bathrooms. . . . My husband and I returned to the States, and after a time, which involved milking three hundred cows in Loveland, Colorado, and being the beekeeper/technician for Dr. Chip Taylor and his grad students, Mark Winston and Gard Otis, in Venezuela for six months, I entered grad school, mostly to have a reason to get out of the house."

Spivak was opposed to being an academic, she says, so she "didn't enter graduate school with the intent of being a professor someday." Then: "After several years and the birth of our son, we decided to move back to Peru. I took a leave of absence from graduate school. We moved back when Bryan was fourteen months old and stayed another year."

Upon returning to the States, Spivak learned she had received a grant from the National Science Foundation to do research on Africanized bees in Peru, but she chose to complete the grant work in Costa Rica.

The grant did not pay enough to live on, so Spivak worked for the famous researcher of social insects Mary Jane West-Eberhard, whose work Gro Amdam had researched when traveling to Brazil to work with Zilá Simões. Spivak entered references and transferred data to the computer, measured spider webs, and performed other tasks. These funds enabled her to pay a Costa Rican couple to help her with the African honey bees.

In this environment, Spivak confirmed data that she had learned when in Peru: "African honey bees are much better adapted to the tropics than are European honey bees. European bees need to be fed to stay alive; they are weak there. African honey bees know how to 'live for the day' and get by on just enough stores in the nest. They just out-compete European bees, but don't destroy apiaries. When queens mate, they encounter African drones, and over time, approximately three years, the population turns African. In some cases, a swarm will invade a very weak, almost dying European colony, but I didn't see that happen much."

Since her work was more flexible than a conventional forty-hour week in the United States, Spivak found it much easier to network to arrange for childcare for her son. "It was much easier being a single parent in Latin America than it was in the U.S.," she says. "Women naturally network to take care of children there—maids, mothers, grandmas, older sisters—all very naturally take care of kids in the neighborhood. Having a child and working is normal and supported by this network. In the U.S., I had to pay lots of money to find safe places for my kid to meet other kids, primarily to day care centers, while I was in school. It takes a village is absolutely correct and we don't have that kind of village support for kids and mothers in the U.S."

Venezuela

The Portuguese explorer Christopher Columbus once swore that Venezuela was the original Eden. To hear Anita Collins talk about it, Venezuela, located at the top of South America, could very well still be. As a newly minted researcher with the US Department of Agriculture bee lab in Baton Rouge, Collins trekked back and forth to Venezuela for ten years, from 1979 to 1989, with a team of researchers to study the effects of the African honey bee. In this time span, the US and Venezuelan teams developed a queen matrix, studied drone congregations, focused on pheromones and foraging behaviors, and determined likelihoods for hygienic and gentleness alleles. The goal was to develop strategies that would assist North American beekeepers when the African honey bee finally arrived in the 1990s.[29]

As a behavioral geneticist, Collins was hired to work on colony defensiveness. In the ten years she was with the team, she was able to resolve two things: a relatively reliable test to assess and quantify colony defensiveness, and an understanding that beekeepers could not select for gentleness with African honey bees. Bee breeders would need to maintain requeening programs in an effort to keep their colonies less defensive.

But these conclusions belie the intense logistics that went into the study. In fact, by the time Collins left the Baton Rouge lab to go to the South Texas Weslaco lab, the team still had not finished its work. The Baton Rouge team made two two-month trips to Venezuela. One was made before Christmas, which in Venezuela was the dry season when the team could really work bees. The other was made after Christmas, during the rainy season. "You couldn't really do anything during that season," Collins says. "It was like being in a shower."

Venezuela had more hybrid African honey bees than strains elsewhere, such as Mexico, and much effort went into controlling drone saturation areas; but the conclusion was still the same: African honey bees were too defensive for North American beekeepers.

That being said, Collins's experiences with people were quite the opposite from her experiences with African honey bees. In a country known for its machismo, Collins encountered very little negative discrimination. Because of her pale skin, which would blush in the hundred-degree heat, Collins would often receive kind treatment by clerks determined to make sure she had access to fans. At security checkpoints, guards would wave her through, while her male colleagues often had to show passports, open hives, and submit to other inconveniences associated with military order. This special treatment did not transfer to Mexico when Collins was there, however. "In Mexico, the people were much more accustomed to women in power," she said. Nonetheless, Collins's experience in Venezuela imprinted on her more than just biology of African honey bees. "The birds were just incredible: the macaws, the oilbirds that live like bats. Even beyond the biology aspects, which were just amazing, I would have never gone to these places without being a beekeeper!"

Bolivia

Bolivia is located in the center of South America. One of the more recent efforts to provide beekeeping extension has had mixed success given the political volatility in the region. In 2000 the municipal government of Villa Vaca Guzman gave land to create a center to produce jams and organic honey, with support provided by the British Embassy. The nongovernmental organization (NGO) Fundacio Chaco provided further project-based assistance, with the goal of providing eleven different communities at least two beehives. The communities ranged in size from a few dozen residents to a few hundred. The largest community, Villa Vaca Guzman (Muyupampa), had approximately twenty-five hundred residents. A Peace Corps volunteer named Conor Branch reported that every single one of the beehives collapsed, so during his brief time there, the communities approached the project from a different perspective. The community villagers would find feral colonies (queens) and then transfer the colonies to conventional box hives. Usually, the men went to search for bees, often having to cut trees down to find the colonies. Women worked on the value-added products and provided the capacity-building components of the project. Branch and his colleagues would visit the community representatives, check on the bees' progress, and discuss the logistics of getting products such as propolis, tinctures, honey, and shampoos to a larger market.

Bolivian women bottling honey. Courtesy of Christine Lehner.

"The single largest issue was transportation," explains Branch. "The community started a small store, but it was never able to grow beyond that. It was never able to find an adequate market."[30] Roads could be closed for six weeks at a time. Landslides could be a major problem. Other logistical challenges contributed to the transportation difficulties. Branch noted that the project was spread over a large area, making it difficult to visit the people and the bees often. They could investigate problems only once or twice a month.

The social dynamics of the women also suggested lessons for future NGO projects. Approximately 140 women participated in the cooperative, but there were varying levels of commitment and participation. Some women lost interest, whereas others were very enthusiastic. Their efforts were hampered by the less enthusiastic members of the group. "Group 'buy-in' is a critical component for success," sums up Branch.[31]

In terms of floral resources, though, Branch was quite enthusiastic about the potential for forest-based beekeeping. Uncultivated citrus trees were abundant. The real challenges seemed to be cultural, and the consequence was a sense of helplessness. Working with smaller groups of more motivated associates could make future projects far more successful, Branch felt. It also seemed to Branch that long project timelines and frequent support and follow-up are essential to project success.

Although in the twentieth century South America was often called the "continent of tomorrow," contemporary women beekeepers do not have a mañana approach to their work. The continent blends research that goes back to the Cretaceous period with a new economics based on organic agricultural production and queen production. In serving as a stepping-stone for women bee scientists, it has provided the world a much better understanding of evolutionary keys, such as nutrition, evolution, foraging behavior, pheromones, and sociality.

Conclusion

Counting for Nothing

Kairos: a passing instant when an opening appears which must be driven through with force if success is to be achieved.
—E. C. White

The Greeks once worshipped two gods of time: Chronos and Kairos. Chronos came to be associated with linear, measurable time; the word "chronology" is our best-known, most widely used etymological reference to this god. For most of us, chronological time is all we have ever known. It was the only form of time taught in my school. As I negotiated a career, I followed its "forced march" toward academic success: college, a graduate program, a doctorate. Its rhythm was a regulated walk away from my farming background, unpredictable weather patterns, and uninsulated farmhouses.

This march came to a screeching halt in 1997 when I offered to help my grandfather with his bees. I would be available one day only, I dictated. Memories of being hot, dusty, and sunburned while adults discussed things they could not control, such as weather, market prices, and loan rates, still lingered. I was done with those days.

My self-righteousness faded as every single thing became a struggle for my grandfather that morning: putting on gloves, fooling with zippers, lighting the smoker. I remember the overwhelming sadness of watching my grandfather's once adept hands tremble constantly. The control I had asserted began to ebb away incrementally. As we lifted a beehive cover and I peered into the frames below me, the steady drumbeat of Chronos stopped, permanently replaced by the softer and steadier hum of honey bees.

Kairos we hear less about. I was in my thirties, with a couple of failed attempts at tenure tracks and conventional relationships behind me, before I learned about Kairos. Kairos can be either gender, although the Greeks first created an icon that was male. The signature feature for the icon in both genders has been that the forehead is covered with hair and the back of the head completely bald. The point: one must grasp the opportunity in front of oneself; otherwise, it will be gone forever, depicted by baldness. Sometimes Kairos is depicted as running by, emphasizing that opportunities are fleeting and will not wait. Not surprisingly, Christian theologians appropriated Kairos to mean a time of divine revelation, and some of the finest writing about it has come from twentieth-century Christian philosophers.

Ten years after that moment in the beehive, I was invited to participate in an environmental tour of surface-mine sites. Acres and acres of nonnative fescues and invasive trees covered land that once had been dense, diverse forests. For me, it was a moment of Kairos, of looking at the acres of fescues on compacted land and deciding, "We can do better than this."

There was just one problem: the classic conflict between Chronos and Kairos. How to balance my new interest in pollinator reclamation with my career as an English professor? Many people would give anything to teach at a private, liberal arts college with small class enrollments. But I had been on one-year contracts for ten years. The teaching schedule given to me at the last minute each summer was an entirely new set of classes.

Seeing an opportunity in front of me, I tucked my hair into a baseball cap and arranged for a flight to Mackay, Australia. A sugar industry town in the nineteenth century, Mackay is now a port city for Peabody Coal. Looking down from my window seat in the plane,

I could see Chinese ships floating like an armada in a harbor, waiting to be loaded with coal as far as I could see.

When I returned from Australia, several forces aligned to create Coal Country Beeworks, a cooperative extension project that works with coal companies to plant pollinator-friendly vegetation and trees. One coal company decided to reclaim surface-mine sites with pollinator-friendly plants for ten years, and two beekeepers decided to sponsor the project. A beekeeping couple in Tennessee, Elaine and Edwin Holcombe, provided the project with trachea mite–resistant bees. By January 2008 a new form of time had started for me: bee time.

I want to create a honey corridor between Kentucky and West Virginia in which landscapes are reclaimed with pollinator-friendly plants. Opportunities for a new economy are in front of us just as they were for the women profiled in this book. Contemporary economies tend to devalue women and the services they provide—whether it be childcare, retail, education, or agriculture. But women's services were not always held in low esteem, nor have women always gone along with the larger market economies. They have found resolutions; they have created new systems or sometimes resurrected ancient ones. So we can change current and flawed economies with new rules, ones that women can construct if we seize the opportunities in front of us.

Former prime minister Marilyn Waring is the inspiration for this transition. In 1975 Waring was elected to parliament in New Zealand at the age of twenty-two. She was reelected three times and was instrumental to making New Zealand a nuclear-free zone. This service was her education in capitalism and economics. She demystifies the language of economics. In the 1995 documentary *Who's Counting: Marilyn Waring on Sex, Lies and Global Economics,* she explains it used to be that "'value' meant to be strong, worthy," but her time in office taught her that the word had been co-opted to mean numbers and accounts. Those who have the most value, Waring discovered, were often invisible in the account books. Women, children, the environment—economics just did not have a value for them.

Economic systems include everything with a cash-generating capacity but recognize "no value other than money. No value to peace, no value to the preservation of natural resources, no value of unpaid work including the value of reproduction and day care." Waring says, quite simply, "As long as activity is going on in [the] market, it is fine. . . . Fantastic growth is aligned with Exxon tragedy . . . insurance, compensations, etc."

More pointedly for this book, Waring learned the economic rules were not specific to New Zealand: "These are the rules everywhere. . . . An African woman walks five miles to get water and tends goats and cattle and children. The men, they worry."

Economics, Waring concludes, "is a tool of people in power . . . a justification."

Currently, industrial agriculture and even industrial apiculture do not "count" women beekeepers in the market economy, which measures profits and bottom lines. But it does not have to be this way.

After three years of the Coal Country Beeworks project, more coal companies are voluntarily planting pollinator-friendly trees and wildflowers. The beeyard sites are open to local communities for workshops. Women teach at and attend the local bee schools as much as men. More acres, more flowers, and more bees than I can possibly count are the beneficiaries of this new economy.

Creating a new economy is not easy. Cultural challenges such as religion and politics have to be negotiated. Even when women attain higher education, their degrees do not level the pay threshold. Similar approaches will not be possible in the short term for every continent. However, the time is right for a new economy — and perhaps a new form of time — in which women and environment count. Counting for nothing gets old after a while.

Notes

Introduction

1. Neumann and Carreck, "Honey Bee Colony Losses."
2. Wollstonecraft, *Vindication of the Rights of Woman*, 305.
3. Davis, "Equal Pay Should Be a Reality, Not a Dream," A3, A7.
4. Gerald, "Black Writer and His Role," 370.
5. Showler, "Some Bee Books by Women Writers: Part Three," 23.
6. Sipe, "How Picture Books Work," 101.
7. Berry, "Queen Production."
8. Wordsworth, "Sonnet 33," 206.
9. Margulis, *Symbiotic Planet*, 151.

Africa

1. Winston, *Biology of the Honey Bee*, 9.
2. Ransome, *Sacred Bee in Ancient Times and Folklore*, 28.
3. Crane, *World History of Beekeeping and Honey Hunting*, 170, 602.
4. Slackman, "Stifled, Egypt's Young Turns to Islamic Fervor," A1.
5. Slackman, "Stifled, Egypt's Young Turns to Islamic Fervor."
6. Bradbear, "Importance of Honey in Third World Countries."
7. Brittan, "Introduction to Modern Beekeeping in Libya (Cyrenica), Part Two," 5.
8. Brittan, "Introduction to Modern Beekeeping in Libya (Cyrenica), Part One," 145.
9. Brittan, "Introduction to Modern Beekeeping in Libya (Cyrenica), Part One," 146.

10. Brittan, "Introduction to Modern Beekeeping in Libya (Cyrenica), Part One," 146.

11. Wakefield and Cook, "Sting in the Tale."

12. Wakefield and Cook, "Sting in the Tale."

13. Crane, *Making a Beeline*, 189.

14. McGregor, "Forty Years as a Beekeeper in South Africa."

15. Crane, *Making a Beeline*, 236.

16. Guy, "Commercial Beekeeper's Approach to the Use of Primitive Hives," 20.

17. Crane, "Beekeeping Techniques," 61.

18. Crane, "Beekeeping Techniques," 61.

19. Crane, "Beekeeping Techniques," 61.

20. Crane, *Making a Beeline*, 190.

21. Crane, *Making a Beeline*, 190.

22. Mickels-Kokwe, *Small-Scale Woodland-Based Enterprises with Outstanding Potential*, 47.

23. Mickels-Kokwe, "Zambia Beekeeping."

24. Mickels-Kokwe, "Zambia Beekeeping."

25. Mickels-Kokwe, *Small-Scale Woodland-Based Enterprises with Outstanding Potential*, 44.

26. Mickels-Kokwe, "Zambia Beekeeping."

27. Mickels-Kokwe, *Small-Scale Woodland-Based Enterprises with Outstanding Potential*, 50.

28. Mickels-Kokwe, *Small-Scale Woodland-Based Enterprises with Outstanding Potential*, 64.

29. Brokensha, Mwaniki, and Riley, "Beekeeping in the Embu District, Kenya," 120.

30. Brokensha, Mwaniki, and Riley, "Beekeeping in the Embu District, Kenya," 116.

31. Kigatiira, "Bees and Beekeeping in Kenya," 76.

32. Adey, "Nairobi Conference on Tropical Apiculture," 56.

33. Brokensha, Mwaniki, and Riley, "Beekeeping in the Embu District, Kenya," 117.

34. Brokensha, Mwaniki, and Riley, "Beekeeping in the Embu District, Kenya," 117.

35. Crane, *Making a Beeline*, 65.

36. Crane, *Making a Beeline*, 65.

37. Kessler, *Best Beekeeper in Lalibela*.

38. Kessler, *Best Beekeeper in Lalibela*.

39. Barrett, *Immigrant Bees 1788–1898*, 2:127.

40. Attridge, *Beekeeping in South Africa,* 181.

41. Mathie, "Woman Beekeeper 'Sees It Through,'" 27–28.

42. Cooper, "Bees Also Need Good Management," 2030.

43. McGregor, "Forty Years as a Beekeeper in South Africa."

44. Mathie, "Woman Beekeeper 'Sees It Through,'" 28.

45. Bressler, "They Roam the Transvaal and Turn Honey into Money," 25.

46. McGregor, "Forty Years as a Beekeeper in South Africa."

47. Crane, *Making a Beeline,* 92.

48. Crane, *Making a Beeline,* 89.

49. Chisnall, "Fungwe."

India

1. Sharma and Nath, *Honey Trails in the Blue Mountains,* 262.

2. Sharma and Nath, *Honey Trails in the Blue Mountains,* 220.

3. Sharma and Nath, *Honey Trails in the Blue Mountains,* 74.

4. Sharma and Nath, *Honey Trails in the Blue Mountains,* 75.

5. Sharma and Nath, *Honey Trails in the Blue Mountains,* 167. *Maanthakula* means honey- or sweetness-marked. Apparently the Badaga people had a talent for gathering honey without destroying the combs, whereas other tribes were called honey cutters because the nests were destroyed in the process. The term was given to the Badagas by Lingayat priests.

6. Sharma, "Women and Bees."

7. Sharma and Nath, *Honey Trails in the Blue Mountains,* 108.

8. Crane, *World History of Beekeeping and Honey Hunting,* 596.

9. Fife, "Concept of the Sacredness of Bees, Honey and Wax in Christian Popular Tradition," 53.

10. Sehgal, *Hindu Marriage and Its Immortal Traditions.*

11. Fife, "Concept of the Sacredness of Bees, Honey and Wax in Christian Popular Tradition," 54.

12. Oldroyd and Wongsiri, *Asian Honey Bees,* 225.

13. Oldroyd and Wongsiri, *Asian Honey Bees,* 210.

14. Williams, *Deforesting the Earth,* 362.

15. Williams, *Deforesting the Earth,* 366.

16. J. C. Douglas, *Handbook of Beekeeping for India,* 6.

17. J. C. Douglas, *Handbook of Beekeeping for India,* 135.

18. J. C. Douglas, *Handbook of Beekeeping for India,* 139.

19. Fletcher, *Beekeeping,* 139.

20. Islam, "Beekeeping in Bangladesh," 271.

21. Bradbear and Reddy, "Sustainable Beekeeping Development in Karnataka," 268–69.

22. Bradbear and Reddy, "Sustainable Beekeeping Development in Karnataka," 267.

23. Bradbear and Reddy, "Sustainable Beekeeping Development in Karnataka," 270.

24. Levi, "Travels in India and Nepal."

25. Sharma, "Women and Bees."

26. "Honey Bees to Bring Money for Women."

27. Sharma, "Women and Bees."

28. Carlson, "My Trip to India."

29. Sharma and Nath, *Honey Trails in the Blue Mountains,* 109.

30. Ede, *Honey and Dust,* 260–70.

31. Shiva, *Stolen Harvest,* 118.

32. Tudge, *Time before History,* 70.

33. Sharma and Nath, *Honey Trails in the Blue Mountains,* 25.

34. Sharma and Nath, *Honey Trails in the Blue Mountains,* 25.

Asia

1. Reid, *Confucius Lives Next Door,* 228–29.

2. Oldroyd and Wongsiri, *Asian Honey Bees,* 253.

3. Oldroyd and Wongsiri, *Asian Honey Bees,* 215–16.

4. Buchmann and Cohn, *The Bee Tree.* In the Malaysian language, the proverb reads, "Tak kenal, tak cinta."

5. C. Crozier, "Overview of Life as a Researcher."

6. R. Crozier, "Overview of Life as a Researcher."

7. Sammataro, "Research Past and Present."

8. De Guzman, "Growing Up in the Philippines."

9. Galton, *Survey of a Thousand Years of Beekeeping in Russia,* 33.

10. Ioyrish, *Bees and People,* 189; Prescott, "Russian Free Economic Society," 508. According to Prescott, the Free Economic Society existed from 1765 to 1917, and in terms of topics of overall importance, beekeeping was fifth, following crop production, estate management, animal fibers, and timber.

11. Ioyrish, *Bees and People.*

12. Ioyrish, *Bees and People,* 199.

13. Glushkov, "Facts about Beekeeping in the U.S.S.R.," 169.

14. Ioyrish, *Bees and People.*

15. Galton, *Survey of a Thousand Years of Beekeeping in Russia,* 50.

16. Glushkov, "Facts about Beekeeping in the U.S.S.R.," 172.

17. Glushkov, "Facts about Beekeeping in the U.S.S.R. (Continued)," 201.

18. Glushkov, "Facts about Beekeeping in the U.S.S.R.," 169.

19. Glushkov, "Facts about Beekeeping in the U.S.S.R. (Continued)," 201.

20. Galton, *Survey of a Thousand Years of Beekeeping in Russia,* 54.

21. Crane, *Making a Beeline,* 28.

22. Williams, *Deforesting the Earth,* 416.

23. Khalifman, *Bees,* 365.

24. Galton, *Survey of a Thousand Years of Beekeeping in Russia,* 54.

25. Anderson, "Beekeeping in the Caucasus," 558.

26. Anderson, "Beekeeping in the Caucasus," 558.

27. Anderson, "Beekeeping in the Caucasus," 558.

28. Ransome, *Sacred Bee in Ancient Times and Folklore,* 155–56.

29. Anderson, "Beekeeping in the Caucasus," 559–60.

30. Anderson, "Beekeeping in the Caucasus," 561, 562.

31. Anderson, "Beekeeping in the Caucasus," 563. It is impossible to tell in the context of Anderson's article if the Russian Imperial Agricultural Society is the same as the Russian Free Economic Society established by Empress Catherine II.

32. Strong, *The Stalin Era,* 48–51.

33. Embrey and Lord, "Turkman Honey Tale," 4–5.

34. Embrey and Lord, "Turkman Honey Tale," 5.

35. Jasim, "Beekeeping in Iraq," 414.

36. Satrapi, *Persepolis 2,* 162.

37. Soza, "The Buzz in Iraq," 30.

38. Hughes, "How Sweet It Is."

39. Hughes, "How Sweet It Is."

40. Sloan, "Afghan Businesswoman Meets U.S. Style," B9.

41. Ransome, *Sacred Bee in Ancient Times and Folklore,* 58.

42. Fife, "Concept of the Sacredness of Bees, Honey and Wax in Christian Popular Tradition," 32.

43. Crane, *World History of Beekeeping and Honey Hunting.*

44. Ransome, *Sacred Bee in Ancient Times and Folklore,* 57.

45. Pundyk, *Honey Spinner,* 234.

46. Pundyk, *Honey Spinner,* 218.

47. Pundyk, *Honey Spinner,* 242.

48. Pundyk, "Continuing the Honey Spinner."

49. Pundyk, "Further Recollections on Turkey."

50. Friedman, "Letter from Istanbul, Part Two."

Europe

1. Klarer, "Women and Arcadia," 13.

2. Fife, "Concept of the Sacredness of Bees, Honey and Wax in Christian Popular Tradition," 16.

3. Fife, "Concept of the Sacredness of Bees, Honey and Wax in Christian Popular Tradition," 24.

4. Williams, *Deforesting the Earth*, 96.

5. Fife, "Concept of the Sacredness of Bees, Honey and Wax in Christian Popular Tradition," 83.

6. Fife, "Concept of the Sacredness of Bees, Honey and Wax in Christian Popular Tradition," 64.

7. Crane and Graham, "Bee Hives of the Ancient World. 1," 34.

8. Crane, *World History of Beekeeping and Honey Hunting*, 193.

9. Fife, "Concept of the Sacredness of Bees, Honey and Wax in Christian Popular Tradition," 126.

10. Fife, "Concept of the Sacredness of Bees, Honey and Wax in Christian Popular Tradition," 126.

11. E.J.P., "Beekeeping for Women," 1.

12. Pantziara, "From Ancient to Modern," 29.

13. Pantziara, "From Ancient to Modern," 29.

14. Marchese, *Honeybee*, 104.

15. Pundyk, *Honey Spinner*, 210.

16. E.J.P., "Beekeeping for Women," 1.

17. Bauman, *Women and Politics in Ancient Rome*, 2.

18. Fife, "Concept of the Sacredness of Bees, Honey and Wax in Christian Popular Tradition," 194–95.

19. Crane and Graham, "Bee Hives of the Ancient World. 1," 29. The differences between Egyptian and Roman hives are discussed more thoroughly in Crane's *World History of Beekeeping and Honey Hunting* (186). Whereas Egyptians and the Hittite cultures used clay cylinders, the Romans differed by making long, rectangular hives from dried fennel stalks. These were called ferula hives and had the advantage of being easily constructed. These ferula hives were still in use during the twentieth century among fourth- and fifth-generation beekeepers.

20. Dadant, "Our Foreign Bee Notes," 177.

21. J. Brown, "Everyday Life, Longevity, and Nuns in Early Modern Florence," 116.

22. Fife, "Concept of the Sacredness of Bees, Honey and Wax in Christian Popular Tradition," 278–79.

23. J. Brown, "Everyday Life, Longevity, and Nuns in Early Modern Florence," 126–27.

24. J. Brown, "Everyday Life, Longevity, and Nuns in Early Modern Florence," 118, 133.

25. Dadant, "Italian Bees," 149.

26. Newman, "Apicultural News Items—Historical Item about Madame De Padua," 355.

27. Newman, "Apicultural News Items—Historical Item about Madame De Padua," 355.

28. R. Brown, *Great Masters of Beekeeping,* 149.

29. Blow, *Beekeeper's Experience in the East,* 33–34.

30. Marchese, "Montalcino."

31. Marchese, "Montalcino."

32. Mutinelli et al., "Honey Bee Colony Losses Italy," 199–200.

33. Pundyk, *Honey Spinner,* 212.

34. Pundyk, *Honey Spinner,* 212.

35. Knowles, *Medieval Religious Houses,* 25.

36. Prete, "Can Females Rule the Hive?," 118–19.

37. Guest, "Ballyvourney and Its Sheela-Na-Gig."

38. Guest, "Ballyvourney and Its Sheela-Na-Gig," 375.

39. Guest, "Ballyvourney and Its Sheela-Na-Gig," 380, 376.

40. Guest, "Ballyvourney and Its Sheela-Na-Gig," 379.

41. Guest, "Ballyvourney and Its Sheela-Na-Gig," 378.

42. Guest, "Ballyvourney and Its Sheela-Na-Gig," 380.

43. Bowe, "Wilhelmina Geddes, Harry Clarke, and Their Part in the Arts and Crafts Movement in Ireland," 58–59.

44. Bowe, "Wilhelmina Geddes, Harry Clarke, and Their Part in the Arts and Crafts Movement in Ireland," 68–69.

45. Bowe, "Wilhelmina Geddes, Harry Clarke, and Their Part in the Arts and Crafts Movement in Ireland," 69, 74. Bowe quotes from P. O. Reeves, "Irish Arts and Crafts," *Studio* 72, no. 295 (1917): 15–22.

46. McLoughlin, "Saint Gobnait."

47. Fife, "Concept of the Sacredness of Bees, Honey and Wax in Christian Popular Tradition," 380.

48. Dadant, "Our Foreign Bee Notes," 177–78.

49. Dadant, "Our Foreign Bee Notes," 177–78.

50. Galton, "Beeswax as an Import in Mediaeval England," 70.

51. Fraser, *History of Beekeeping in Britain,* 20–21.

52. More, *Bee Book,* 86.

53. Fife, "Concept of the Sacredness of Bees, Honey and Wax in Christian Popular Tradition," 352.

54. Fife, "Concept of the Sacredness of Bees, Honey and Wax in Christian Popular Tradition," 352.

55. Monaghan, "Austēja."

56. Ulrich, *Good Wives,* 7.

57. Fraser, *History of Beekeeping in Britain,* 22.

58. Dadant, "Our Foreign Bee Notes," 78.

59. Fraser, *History of Beekeeping in Britain,* 38.

60. Lachman, *From Manor to Market,* 101, 129.

61. Ulrich, *Good Wives,* 6.

62. Jutte, *Poverty and Deviance in Early Modern Europe,* 168.

63. Fraser, *History of Beekeeping in Britain,* 31.

64. Crane, *World History of Beekeeping and Honey Hunting,* 587–88.

65. F.B., *Office of the Good Housewife.*

66. F.B., *Office of the Good Housewife,* A3, 13, 105.

67. F.B., *Office of the Good Housewife,* 20.

68. F.B., *Office of the Good Housewife,* 129–30.

69. Walker, "Bee Boles and Past Beekeeping in Scotland," 112.

70. Warder, *True Amazons,* vi–vii.

71. Warder, *True Amazons,* ix.

72. Warder, *True Amazons,* 21–22.

73. Dummelow, *Wax Chandlers of London,* 98; More, *Bee Book,* 132.

74. More, *Bee Book,* 132, 134.

75. More, *Bee Book,* 133–34.

76. Crane, *World History of Beekeeping and Honey Hunting.*

77. Sweet, "Hibernation and Pollen in 1764," 58.

78. Artman and Hall, *Beauties and Achievements for the Blind,* 87.

79. Dodd, *Beemasters of the Past,* 43.

80. Laidlaw and Page, *Queen Rearing and Bee Breeding,* 11.

81. Artman and Hall, *Beauties and Achievements for the Blind,* 88–89.

82. R. Brown, *Great Masters of Beekeeping,* 36.

83. Johansson and Johansson, "Letter from François Huber," 171–72.

84. Dunbar, "Memoir of Huber," 63.

85. More, *Bee Book,* 126.

86. Runte, "From La Fontaine to Porchat," 84.

87. Beck, "Order of the Honey Bee," 268.

88. Jay, *Extraordinary Exhibitions,* 42. "In 1772, [the beekeeper Daniel] Wildman rode with Astley's Circus. His most famous disciple was Mrs. [Patty] Astley, a woman 'well known for her great Command over the Bees.'"

89. Bagster, *Management of Bees with a Description of the "Ladies' Safety Hive,"* 9, 207.

90. Bagster, *Management of Bees with a Description of the "Ladies' Safety Hive,"* 208.

91. Walker, "Bee Boles and Past Beekeeping in Scotland," 112.

92. Showler, "In the Apiary," 30–31.

93. Crane, "Worshipful Company of Wax Chandlers," 63.

94. Crane, "Century for England," 90.

95. Showler, "In the Apiary."

96. *Instruction in Bee-Keeping for the Use of Irish Beekeepers,* 5–6.

97. Watson, *Bee-Keeping in Ireland.*

98. Watson, *Bee-Keeping in Ireland,* 79.

99. Showler, "Some Bee Books by Women Writers: Part Two 1905–1953," 22–23.

100. Showler, "Some Bee Books by Women Writers: Part Two 1905–1953," 22–23.

101. Showler, "Some Bee Books by Women Writers: Part Two 1905–1953," 22–23.

102. Betts, "Personal," 93.

103. Crane, "Dr. A.Z. Abushâdy," 112.

104. Betts, "Bookshelf: *Archiv Fuer Bienenkunde,*" 57.

105. Betts, "Sackcloth and Ashes," 22.

106. Betts, "Personal," 93.

107. Morgenthaler, "In Memory of Miss Annie D. Betts," 310.

108. Showler, "Some Bee Books by Women Writers: Part Two 1905–1953," 23.

109. Crane, "Bee Laboratory: Research Notes," 5.

110. Crane, "Bumble Bee Honeys and Others."

111. Crane, "Obituaries: Dr. Anna Maurizio, an Appreciation from IBRA," 98.

112. Jones, "Further Notes on Eva Crane's Accomplishments."

113. Jones, "Dr. Eva Crane," 11.

114. Crane, "Bee Laboratory: Research Notes."

115. Crane, "Overview of a Career in Beekeeping."

116. Crane, "Inauguration of the Bee Research Association International Appeal," 121.

117. Bradbear, "Importance of Honey in Third World Countries."

118. Pundyk, *Honey Spinner,* 186–87.

119. Rocheblave, "A French Woman's Experience as a Beekeeper."

120. Showler, "Some Bee Books by Women Writers: Part Two 1905–1953."

121. Knight, "Cath Keay."

122. Knight, "Cath Keay."

123. Amdam, "Overview of Research in Beekeeping."

124. McNeil, "Bee Time."

125. Traynor, "Bee Breeding around the World," 136.

126. McNeil, "Bee Time," 541–42.

North America, Part 1

1. Engel, Hinojosa-Diaz, and Rasnitsyn, "Honey Bee from the Miocene of Nevada."

2. Irving, *Diedrich Knickerbocker's History of New York*, 172–73.

3. Jefferson, *Notes on the State of Virginia*, 71–72.

4. Ulrich, *Good Wives*, 6.

5. Kupperman, "Beehive as a Model for Colonial Design," 281. See also Kupperman, "Death and Apathy in Early Jamestown" for more extensive research into low morale in male-only colonial efforts.

6. Kupperman, "Beehive," 284.

7. Kupperman, "Beehive," 287–88.

8. Ulrich, *Good Wives*.

9. Withington, "Republican Bees."

10. Waldstreicher, *Runaway America*, 176.

11. Earle, *Diary of Anna Green Winslow*, 108.

12. Biddle, *Extracts from the Journal of Elizabeth Sandwith Drinker, from 1759 to 1807, A.D.*, 117.

13. Forman, *Diary of Elizabeth Drinker*, 169, 845.

14. Nash, "Mrs. Ballard's Diary, 1785–1812," 313.

15. Withington, "Republican Bees."

16. E.J.P. "Beekeeping for Women," 3.

17. In a bee journal written in 1851–1854, "A Journal Concerning Bees in the Second Order," Shaker Giles A. Avery recorded the daily trials and tribulations of working with bees and bee swarms in Lebanon, New York.

18. Patterson, *Shaker Spiritual*, 427–28.

19. Traill, *Backwoods of Canada*, 351, 135.

20. Traill, *Backwoods of Canada*, 188.

21. Griffith, "On Bees." This review was written by Griffith anonymously, but her identity was made known in Hale, "Literary Notices." Thacher wrote admiringly about the Charlieshope hive, calling Griffith an "inventress" in *A Practical Treatise on the Management of Bees* (93).

22. Griffith, "On Bees," 348–49.

23. Kritsky, *Quest for the Perfect Hive,* 85.

24. Griffith, "On Bees," 346.

25. Seaton, "Mary Griffith," 148.

26. Flottum, *Backyard Beekeeper.*

27. Crane, *World History of Beekeeping and Honey Hunting,* 423.

28. Langstroth, "Langstroth's Reminiscences: How He Became Interested in Bees," 80. This article was taken from a larger unpublished manuscript Langstroth wrote, "Langstroth's Reminiscences," housed at Cornell Library, Phillips Collection. Several other articles were taken from this manuscript and published in *Bee Culture* in intermittent issues.

29. Langstroth, "Langstroth's Reminiscences: Getting the Movable Frame Introduced, etc.," 206.

30. Langstroth, "Hive and the Honey Bee," 92.

31. Langstroth, "Letter to Charles Dadant." Although we may never know the full details of their marriage, Langstroth continued to struggle financially after Anna's death. In correspondence with Charles Dadant in the late 1880s, in which discussion of future publication of *The Hive and the Honey Bee* was the primary topic, the details of the Langstroth poverty were better defined. In a letter dated October 5, 1887, Langstroth details more tellingly the desperate circumstances in which he had been living since Anna's death: "I have suffered much for the last two years from our means not permitting us to procure the nourishing diet needed by one of my age and infirmities. We have not meat on our table on an average more than once a week, and not enough fruit. . . . We have been able only to buy what would satisfy hunger at the least cost. We have preferred to suffer rather than make known our wants to our most generous friends." In a letter dated May 10, 1888, Langstroth states the anxieties that his requests for assistance may have had on his friendship with the Dadants: "Your loans of money came to me at a time when I was surely in need of help, but I would much rather that my book should forever have gone out of print, than to have knowingly done anything to damage the interests of friends whom I respect and love."

In spite of the stress and financial difficulties, Langstroth remained optimistic. In a letter dated February 2, 1888, he concludes a note to Dadant by praising the "tongues in trees, books in running brooks, / Sermons in bees and God in everything." And he never lost his affection for his wife. In a letter dated November 19, 1888, he mentions an article written by C. C. Miller just published in *Bee Culture*: "I would be particularly pleased if you should see fit to do so, not only because it is fuller, but specially because of what is said of my dear wife."

32. Hoffman, "Eureka!," 33.

33. Dumas, "Apiculture in Early Texas."

34. Dumas, "Apiculture in Early Texas."

35. Dumas, "Apiculture in Early Texas," 120.

36. Dumas, "Apiculture in Early Texas," 117.

37. B. Weaver, "History of the Zach Weaver Family and Weaver Apiaries," 568.

38. The most famous of these narratives was about a woman named Mary Jemison, who was captured by the Seneca tribe, left, and then went back to marry a Seneca man and stay with the tribe. See Seaver, *Narrative of the Life of Mrs. Mary Jemison*. Much later, in nineteenth-century Texas, Cynthia Ann Parker was taken captive by Comanche, married into the tribe, and chose to stay. Jackson, *Comanche Moon*, is about this captivity. Coontz, *The Way We Never Were*, and Kupperman, *Facing Off*, describe women's equity within Native American family structures.

39. Dolbeare, *Narrative of the Captivity and Suffering of Dolly Webster among the Comanche Indians in Texas*, 18.

40. Dolbeare, *Narrative of the Captivity and Suffering of Dolly Webster among the Comanche Indians in Texas*, 21.

41. Dolbeare, *Narrative of the Captivity and Suffering of Dolly Webster among the Comanche Indians in Texas*, 22.

42. Oberste, *Texas Irish Empresarios and Their Colonies*, 144–45.

43. Tupper, "Journal." By marrying into this family, Ellen Tupper gained an impressive mother-in-law, Mary Ann McIlwayne Tupper. Mary Ann arrived in America from Ireland and married Jonathan Tupper, who died from yellow fever almost immediately after Allen was born. A widow, Mary Ann started her own business as a milliner in Bath, Maine. She later moved to Bangor when the business expanded to employ eight women year-round and twice that number during busy seasons. In addition, Mary Ann Tupper was an abolitionist, a temperance worker, and a "friend to the fallen," as Ellen Tupper reminisced diplomatically. "Long before I knew your father," Ellen wrote to her daughter Margaret, "I knew his mother by reputation, and what first attracted me to him was the curiosity to see the son of such a mother. In those days, a business woman was seldom seen" (3).

44. True and Kirby, *Allen Tupper True*, 8.

45. Hanaford, *Daughters of America*, 703–4.

46. Tupper, "Beekeeping," 152.

47. Fisher, *In the Beginning, There Was Land*, 342–43.

48. Newman, "Queen Bee's Temptation," 53.

49. Fisher, *In the Beginning, There Was Land*, 343.

50. Root, "Headlines—Ellen Tupper," 54.

51. Dadant, "Down with the Importation of Bees," 55.

52. Tupper, "Answer to Mr. Dadant," 51.

53. Root, "Honor Roll—Ellen Tupper," 5.

54. Root, "Humbugs and Swindlers—Letter Written by Will R. King," 93.

55. True and Kirby, *Allen Tupper True,* 8.

56. Newman, "Notes and Queries—Ellen Tupper," 85.

57. "Forgeries by a Woman."

58. Root, "Notes and Queries—Letter from William Ellsworth," 117.

59. Newman, "Mrs. Tupper's Troubles," 252.

60. Newman, "Notes and Queries," 120.

61. Hanaford, *Daughters of America,* 703–4.

62. Newman, "Ellen Tupper—Obituary," 179.

63. Willard and Livermore, *American Women,* 726.

64. Working, *Bees in Colorado.*

65. Newman, "Notes and Queries—Mrs. F.A. Dunham's Wax Foundation," 41.

66. Marshall and Marshall, "Letter."

67. Whitney, "Bees in California," 19.

68. Root, "Roll Call: Miss Annie C. Mann," 6.

69. Newman, "Notes and Queries—Spaid Honey House," 56.

70. Baldridge, "Adulteration of Honey," 181–82.

71. Baldridge, "Amende Honorable—Errata," 206–7.

72. Crockett, " 'Beekeeping for Women.'—Read at Maine Convention," 473.

73. Ellis, "Early History of the Nebraska Beekeepers Association: Mrs. J. N. Heater." See also Newman, "Mrs. J. N. Heater," 209.

74. Harrison, "Lady Presidents," 58.

75. Harrison, "Maddened Bees-Robbing," 245.

76. Harrison, "Excelsior—A Lady's Experience," 295.

77. Harrison, "Watch the Swarms," 407.

78. Keyes, "Letters in Our Box."

79. Dunham, "Langstroth."

80. Newman, "Lady Beekeepers," 523.

81. Harrison, "International Congress," 35.

82. Linswick, "Nellie's Experiment," 274–75.

83. Linswick, "Shall Women Keep Bees.—Read before the Michigan Convention," 170.

84. Bills, "Can Women Keep Bees with Profit"; Baker, "A Woman's Experience.—Read before the Michigan Association," 194–95.

85. Phelps, "Women as Beekeepers," 315.

86. Phelps, "Women as Beekeepers," 315.

87. Baker, "What Shall We Wear?," 50.

88. Stover, "Visiting California's Apiaries," 219.

89. Fauls, "Voices from the Hives," 21.

90. Chaddock, "Beekeeping for Women," 422.

91. Roberts, "Ann Gudgel," 28.

92. Chaddock, "Mrs. Chaddock Talks to Us about Market Gardening and Bees," 480–81.

93. Stowe, "Some Kindly Hints to Young Women," 20.

94. Newman, "Canadian Apiarists in London," 728.

95. Clarke, "What Constitutes a Good Bee-Periodical?"

96. Root, "Humbugs and Swindlers—Lizzie Cotton," 41.

97. Newman, "Lizzie Cotton," 147.

98. Cotton, *New System of Beekeeping*, 11–12.

99. Cotton, *New System of Beekeeping*, 20.

100. Cotton, *New System of Beekeeping*, 132–33.

101. Phillips, "Beekeeping in the Post-war Era," 49.

102. Root, "Notes and Queries—A Kind Word for Mrs. Cotton," 588.

103. Savery, "Experiences of a Beekeeper."

North America, Part 2

1. Stoll, *Fruits of Natural Advantage*.

2. Ferrara, *Love Affair with Nature*, 29.

3. A. Comstock, *Comstocks of Cornell*; Henson, "Comstock Research School in Evolutionary Entomology." Anna Comstock's reputation attracted a young high-school student named Edith Marion Patch. With prize money she won during high school, Patch bought a copy of the Comstocks' *Manual of Insects* and became determined to study with them one day. Patch's dream was deferred until she was established at Maine Agricultural Experiment Station. While maintaining her research and extension duties in Maine, she worked on a master's degree at Cornell, benefiting greatly from the evolutionary systematics and fellowship provided by both John and Anna Botsford Comstock. She was one of the first to predict that entomologists may have to protect insects instead of eradicating them. Patch became the first woman president of the Entomological Society of America. Bird, "Dame Bug and Her Students."

4. Vogt, "Anna Botsford Comstock," 50.

5. J. H. Comstock, *Introduction to Entomology*.

6. A. Comstock, "Beekeeping for Women," 51–52. Emphasis original.

7. B. Washington, *Tuskegee and Its People*.

8. M. Washington, "We Must Have a Cleaner Social Morality."

9. Phillips, "Beekeeping in the Post-war Era," 49.

10. McConnell, "Personal Experience of a Teacher and Beekeeper," 49.

11. Pellett, "How Women Win," 372–73.

12. York, "Miss Emma Wilson's Beekeeping Sisters Department." See also Wilson, "Our Bee-Keeping Sisters"; and Wilson, "Bee-Keeping for Women."

13. Pellett, "Mrs. C. C. Miller and Miss Wilson Die," 169.

14. Pellett, "How Women Win," 372.

15. Allen, "Dignity of Beekeeping," 10.

16. Noel, "My Bees," 25.

17. Pellett, "Honey Production in the Sage District," 296.

18. Stewart, "Uncle Josh and the Honey Bees."

19. Kerr, "My Life."

20. Kerr, "My Life," 10.

21. Phillips, "Beekeeping in the Post-war Era."

22. Geithmann, "Story of Lillian Hill, Beekeeper of the Elbe," 167–68.

23. Miller, *Sweet Journey*, 84.

24. Quoted in Peiss, *Hope in a Jar*, 135.

25. Peiss, *Hope in a Jar*, 135.

26. Peiss, *Hope in a Jar*, 197.

27. Peiss, *Hope in a Jar*, 240–45.

28. Phillips, "War Time Regulations Affecting the Beekeeping Business."

29. Phillips, "War Time Regulations Affecting the Beekeeping Business."

30. Glatz, "My Life as a Beekeeper."

31. Wien, Glatz, and Morse, "Flowering, Pollination, and Fruit Set."

32. Glatz, "My Life as a Beekeeper."

33. See "Civil Rights Act of 1964."

34. Collins, "Career as a Researcher in Federal Bee Laboratories."

35. Collins, "Conversation about Being a Beekeeper."

36. Collins, "Genetic Diversity."

37. Sammataro, "Family Beekeeping History."

38. Sammataro, "Research Past and Present."

39. DiGrandi-Hoffman, Sammataro, and Alarcon, "Importance of Microbes in Nutrition and Health of Honey Bee Colonies."

40. Sammataro, "Research Past and Present."

41. De Guzman, "Growing Up in the Philippines."

42. DiGrandi-Hoffman, "Career as a Researcher."

43. Frazier, "Experiences as Bee Inspector in US and Africa."

44. Maryann Frazier, "Update on Colony Collapse Disorder in Honey Bee Colonies in the United States," 1. Frazier acknowledges that she borrowed the analogy from Gerald Hayes, Florida state bee inspection director.

45. Spivak, "Social Safety Networks for Families."

46. Spivak, "Overview of Costa Rica Experience."

47. Spivak, "Minnesota Hygienic Bees."

48. Spivak, "Social Safety Networks for Families."

49. Spivak, "Coordinated Agricultural Project (CAP) Grant Project," 44.

50. Flottum, "Marla Spivak Wins MacArthur Fellowship," 37.

51. Crane, *Archeology of Beekeeping*.

52. McDonough, "Building a Better Bee."

53. McDonough, "Building a Better Bee."

54. McDonough, "Building a Better Bee"; Cobey, personal comment.

55. McDonough, "Building a Better Bee."

56. Berry, "Career in Bees."

57. Flottum, "Jennifer Berry . . . Queen Producer," 42.

58. Flottum, "Jennifer Berry . . . Queen Producer," 43–44.

59. Flottum, "Jennifer Berry . . . Queen Producer," 43–44.

60. Ostiguy, "New Integrated Pest Management Techniques in Beekeeping."

61. Ostiguy, "Managed Pollinator CAP," 52.

62. Ostiguy, "New Integrated Pest Management Techniques in Beekeeping."

63. Ostiguy, "Career in Bee Research."

64. Cox-Foster, "Choosing a Career in the Sciences."

65. Berenbaum, "Bee Afraid, Bee Very Afraid."

66. Grozinger, "Research with Queen Pheromones."

67. McGaughey, "Grozinger Found Her Inspiration at Illinois."

68. Grozinger, "Research with Queen Pheromones."

69. Park-Burris, "Overview of Park-Burris Family History."

70. Strachan-Severson, "Overview about Strachan Queens."

71. Woodworth, "Migratory Beekeeping Challenges."

72. Tarwater, "Migratory Beekeeping in TN."

73. Tarwater, "Inspecting Bees in Florida."

74. Tarwater, "Inspecting Bees in Florida."

75. Bee, "Writing a How-To Book."

76. Bee, "Gender Differences in Swarm Removal."

77. Bee, "Gender Differences in Swarm Removal."

78. Bee, "Writing a How-To Book."

79. Morandin, "Wild Bees and Agroecosystems."

80. Lehman, "Kids and Bees Programs."

81. Holcombe, "Importance of the Honey Queen Program."

82. Sanford, "Florida Standard of Identity for Honey," 21.

83. Gentry, "National Honey Standard."

84. Graham, "Florida Adopts Honey Identity Standard," 287.

85. Gentry, "National Honey Standard."

86. Phipps, "International Honey Market Report" (147, no. 12), 1016.

87. Pasco, "Transshipped Honey Hurting U.S. Beekeepers."

88. Webb, "World Honey Shows and Honey Standards."

89. Flottum, "Winding Down at the White House."

90. Sanford, "New Generation Honey Cooperatives," 17.

91. O'Brien, "Ontario's Tech Transfer Team," 47.

92. O'Brien, "Ontario's Tech Transfer Team," 49.

93. Tam, "Bee Girls Tech Transfer Team."

Australasia

1. Barrett, *Immigrant Bees 1788–1898: A Cyclopedia*, 76.

2. Barrett, *Immigrant Bees 1788–1898. Volume 2*, 32.

3. Barrett, *Immigrant Bees 1788–1898. Volume 2*, 32.

4. Barrett, *Immigrant Bees 1788–1898. Volume 2*, 94.

5. Barrett, *Immigrant Bees 1788–1898. Volume 2*, 94–95.

6. Apiarian Society of Victoria, *Rules and Bylaws of the Apiarian Society of Victoria*, 2.

7. Benecke, *Commercial Beekeeping in Australia*, 14.

8. Crane, "Letter from Australia," 86.

9. Crane, "Letter from Australia," 86.

10. Tennant, *Honey Flow*, 21.

11. Tennant, *Honey Flow*, 20.

12. Tennant, *Honey Flow*, 1.

13. Tennant, *Honey Flow*, 15.

14. Tennant, *Honey Flow*, 347.

15. Crane, *Making a Beeline*, 67–68, 80.

16. Wells, *Men of the Honey Bee*, 6.

17. Wells, *Men of the Honey Bee*, 192.

18. Wells, *Men of the Honey Bee*, 202.

19. Weiss, "Federal Council of Australian Apiarist's Associations," 501.

20. McAdam, "Being a Beekeeper on Kangaroo Island."

21. Greenwood, "Bees on Their Knees."

22. Bourke, "Cradle Mountain Honey."

23. Campion, "Kiwi Queen Bee," 40.

24. Campion, "Kiwi Queen Bee," 40.

25. Pundyk, *Honey Spinner*, 316–17.

South America

1. Padilla et al., "Bees, Apiculture and the New World," 563–64.

2. Crane, "Continent of Tomorrow," 105.

3. Crane, "Continent of Tomorrow," 105.

4. Sanford, "Africanized Honey Bee," 597–98.

5. Sanford, "Africanized Honey Bee," 598; Crane, "Continent of Tomorrow," 105.

6. Crane, *Making a Beeline,* 160.

7. Crane, "Central African Bee in South America," 117–18.

8. Crane, "Central African Bee in South America," 118.

9. Crane, *Making a Beeline,* 160.

10. Crane, "Central African Bee in South America," 119.

11. Soares, "Building a Future on Science," 85.

12. McNeil, "Bee Time," 540.

13. Amdam, "Overview of Research in Beekeeping."

14. McNeil, "Bee Time," 540.

15. Landim, "EU Honey Legislation."

16. Phipps, "International Honey Market Report" (148, no. 5), 400.

17. Simões, "Career in Research."

18. Graham, "World Honey Market," 781.

19. Braunstein and Verettoni, "Malka Queens."

20. Braunstein and Verettoni, "Malka Queens," 31.

21. Sanford, "Fourth Individual in a Honey Bee Colony," 17.

22. McPherson, "Argentina Couple Learning about Beekeeping in the United States."

23. Verettoni, "Queen Breeding in Argentina."

24. Braunstein, "Overview of Queen Production in Argentina."

25. Braunstein and Verettoni, "Malka Queens," 31.

26. Graham, "World Honey Market," 781.

27. Verettoni, "Queen Breeding in Argentina."

28. Spivak, "Overview of South America Experience."

29. Collins, "Career as a Researcher in Federal Bee Laboratories."

30. Branch, "Beekeeping in Bolivia."

31. Branch, "Beekeeping in Bolivia."

Bibliography

Adey, Margaret. "Nairobi Conference on Tropical Apiculture." *Bee World* 66, no. 2 (1985): 54–58.

Allen, Mrs. Armstrong. "Dignity of Beekeeping." *Victorian Bee Journal* 1, no. 5 (1919): 10.

Amdam, Gro. "Overview of Research in Beekeeping." Interview, August 16, 2008.

Anderson, Alder. "Beekeeping in the Caucasus." *Windsor Magazine* 26, no. 3 (1907): 557–64.

Apiarian Society of Victoria. *Rules and Bylaws of the Apiarian Society of Victoria.* Melbourne: Clarson, Shallard, and Company, 1861.

Artman, William, and L. V. Hall. *Beauties and Achievements for the Blind.* Rochester, NY, 1869.

Attridge, Alfred. *Beekeeping in South Africa: A Book for Beginners.* Cape Town: Specialty, 1917.

Avery, Giles. "A Journal Concerning Bees in the Second Order." Hancock Shaker Village, Pittsfield, MA, 1851–1854. Pleasant Hill (KY) Shaker Museum.

Bagster, Samuel, Jr. *The Management of Bees with a Description of the "Ladies' Safety Hive."* 3rd ed. London: Saunders and Otley, 1834.

Baker, Mrs. L. B. "What Shall We Wear?" *American Bee Journal* 14, no. 2 (1878): 50–51.

Baker, Mrs. L. B. "A Woman's Experience.—Read before the Michigan Association." *American Bee Journal* 13, no. 6 (1877): 194–95.

Baldridge, M. M. "Adulteration of Honey." *American Bee Journal* 11, no. 8 (1875): 181–82.

Baldridge, M. M. "Amende Honorable—Errata." *American Bee Journal* 11, no. 9 (1875): 206–7.

Barrett, Peter. *Immigrant Bees 1788–1898: A Cyclopedia to the Introduction of European Honeybees to Australia and New Zealand.* Springwood, New South Wales: Peter Barrett, 1995.

Barrett, Peter. *Immigrant Bees 1788–1898. Volume 2: An Update on the Introduction of European Honeybees to Australia and New Zealand.* Springwood, New South Wales: Peter Barrett, 1999.

Bauman, Richard. *Women and Politics in Ancient Rome.* London: Routledge, 1982.

Beck, Bodog. "Order of the Honey Bee." *American Bee Journal* 81, no. 6 (1941): 268.

Bee, Cindy. "Gender Differences in Swarm Removal." E-mail, May 5, 2010.

Bee, Cindy. "Writing a How-To Book." E-mail, August 27, 2008.

Beerbohm, Max. "Ibsen's 'Epilogue.'" *Saturday Review,* May 26, 1906, 650–51. Reprinted in *Henrik Ibsen: The Critical Heritage,* ed. Michael Egan, 443–45. Oxford: Taylor and Francis, 1972.

Benecke, Frederick. *Commercial Beekeeping in Australia.* Barton, ACT: Rural Industries Research and Development Corporation, 2007.

Berenbaum, May. "Bee Afraid, Bee Very Afraid." Interview with Steve Mirsky. *Scientific American,* August 14, 2009. http://www.scientificamerican.com/podcast/episode.cfm?id=bee-afraid-bee-very-afraid-09-08-14.

Berry, Jennifer. "A Career in Bees." E-mail, May 10, 2010.

Berry, Jennifer. "Queen Production." Presentation at Eastern Apicultural Society, Murray, Kentucky, August 5, 2008.

Betts, Annie. "Bookshelf: *Archiv Fuer Bienenkunde.*" *Bee World* 30, no. 7 (1949): 57–58.

Betts, Annie. "Personal." *Bee World* 30, no. 12 (1949): 93.

Betts, Annie. "Sackcloth and Ashes." *Bee World* 30, no. 3 (1949): 22.

Biddle, Henry D., ed. *Extracts from the Journal of Elizabeth Sandwith Drinker, from 1759 to 1807, A.D.* Philadelphia: J. B. Lippincott, 1889.

Bills, Mrs. M. A. "Can Women Keep Bees with Profit." *American Bee Journal* 13, no. 6 (1877): 197.

Bird, Mary. "Dame Bug and Her Students: The Science and Environmental Teaching of Edith Marion Patch." PhD dissertation, Harvard University, 2006.

Blow, Thomas. *A Beekeeper's Experience in the East: Among the Queen Raisers in the North of Italy and Carniola.* Welwyn, UK, 1887.

Bourke, Yeonsoon. "Cradle Mountain Honey." E-mails, July 17 and September 30, 2008.

Bowe, Nicola Gordon. "Wilhelmina Geddes, Harry Clarke, and Their Part in the Arts and Crafts Movement in Ireland." *Journal of Decorative and Propaganda Arts* 8 (Spring 1988): 58–79.

Bradbear, Nicola. "Importance of Honey in Third World Countries." E-mail, December 7, 2007.

Bradbear, Nicola, and M. S. Reddy. "Sustainable Beekeeping Development in Karnataka." In *Asian Bees and Beekeeping: Progress of Research and Development,* ed. M. Matsuka, 266–70. Enfield, NH: Science Publishers, 1998.

Branch, Conor. "Beekeeping in Bolivia." Interview, November 10, 2010.

Braunstein, Martin. "An Overview of Queen Production in Argentina." E-mail, September 21, 2008.

Braunstein, Martin, and Sonia Verettoni. "Malka Queens." *Beekeepers Quarterly* 84, no. 7 (2006): 30–35.

Bressler, Ross. "They Roam the Transvaal and Turn Honey into Money." *Farmer's Weekly,* September 16, 1959, 25.

Brittan, Olive. "Introduction to Modern Beekeeping in Libya (Cyrenica), Part One." *Bee Craft* 37, no. 12 (1955): 145–46.

Brittan, Olive. "Introduction to Modern Beekeeping in Libya (Cyrenica), Part Two." *Bee Craft* 38, no. 1 (1956): 4–5.

Brokensha, David, H. Mwaniki, and B. Riley. "Beekeeping in the Embu District, Kenya." *Bee World* 53, no. 3 (1972): 114–23.

Brown, Judith. "Everyday Life, Longevity, and Nuns in Early Modern Florence." In *Renaissance Culture and the Everyday,* ed. Patricia Fumerton and Simon Hunt, 115–38. Philadelphia: University of Pennsylvania Press, 1999.

Brown, Ron. *Great Masters of Beekeeping.* London: Butler and Tanner, 1994.

Buchmann, Stephen, and Diana Cohn. *The Bee Tree.* El Paso, TX: Cinco Puntos, 2007.

Campion, Caroline. "Kiwi Queen Bee." *Saveur* 93, no. 5 (2006): 40.

Carlson, Reyah. "My Trip to India." http://www.reyahbeesness.com.

Chaddock, Mrs. Mahala B. "Beekeeping for Women." *American Bee Journal* 22, no. 27 (1886): 422.

Chaddock, Mrs. Mahala B. "Mrs. Chaddock Talks to Us about Market Gardening and Bees." *Bee Culture* 13, no. 7 (1887): 480–81.

Chisnall, Keith. "Fungwe." Unpublished manuscript, Johannesburg, South Africa, 2006.

"Civil Rights Act of 1964." Pub. L. 88–352, 78 Stat 241.

Clarke, W. F. "What Constitutes a Good Bee-Periodical? Is It the Duty of the Beekeepers to Sustain Any but the Best?" In *Report of the Proceedings of the Twenty-Second Annual Convention.* Albany, NY: New York State Beekeepers Association, 1891. Reprinted in *American Bee Journal* 27, no. 4 (1891): 111.

Collins, Anita. "A Career as a Researcher in Federal Bee Laboratories." E-mail, May 22, 2010.

Collins, Anita. "A Conversation about Being a Beekeeper." Interview, Heartland Apiculture Society conference, Southern Illinois University, Edwardsville, July 7, 2005.

Collins, Anita. "Genetic Diversity." Lecture at Heartland Apiculture Society conference, Southern Illinois University, Edwardsville, July 8, 2005.

Comstock, Anna Botsford. "Beekeeping for Women." In *The ABC and XYZ of Bee Culture,* ed. A. I. Root, 50–52. Medina, OH: A. I. Root, 1908.

Comstock, Anna Botsford. *The Comstocks of Cornell: John Henry Comstock and Anna Botsford Comstock.* Ithaca, NY: Cornell University Press, 1953.

Comstock, J. H. *An Introduction to Entomology, with Many Original Illustrations Drawn and Engraved by Anna Botsford Comstock.* Ithaca, NY: Comstock, 1888.

Connor, Larry. *Bee Sex Essentials.* Kalamazoo, MI: Wicwas, 2008.

Coontz, Stephanie, ed. *The Way We Never Were: American Families and the Nostalgia Trap.* Rev. ed. New York: Basic Books, 2000.

Cooper, F. H. "Bees Also Need Good Management." *Farmer's Weekly,* September 9, 1936, 2030–31.

Cotton, Lizzie. *The New System of Beekeeping.* West Gorham, ME, 1880.

Cox-Foster, Diana. "Choosing a Career in the Sciences." E-mail, November 20, 2010.

Crane, Elaine Forman, ed. *The Diary of Elizabeth Sandwith Drinker.* Vol. 2. Boston, MA: Northeastern University Press, 1991.

Crane, Eva. *Archeology of Beekeeping.* Ithaca, NY: Cornell University Press / International Bee Research Association, 1983.

Crane, Eva. "Beekeeping Techniques." *Bee World* 53, no. 2 (1972): 60–63.

Crane, Eva. "Bee Laboratory: Research Notes." *Bee World* 30, no. 1 (1949): 5.

Crane, Eva. *Bees and Beekeeping: Science, Practice and World Resources.* Ithaca, NY: Cornell University Press, 1990.

Crane, Eva. "Bumble Bee Honeys and Others." *Bee World* 53, no. 1 (1972): 38–39.

Crane, Eva. "The Central African Bee in South America." *Bee World* 52, no. 3 (1971): 116–21.

Crane, Eva. "A Century for England." *Bee World* 55, no. 3 (1974): 90.

Crane, Eva. "Continent of Tomorrow." *Bee World* 53, no. 3 (1972): 105–6.

Crane, Eva. "Dr. A.Z. Abushâdy." *Bee World* 36, no. 6 (1955): 112.

Crane, Eva. "Inauguration of the Bee Research Association International Appeal." *Bee World* 42, no. 5 (1961): 117–23.

Crane, Eva. "Letter from Australia." *Bee World* 48, no. 3 (1967): 86–87.

Crane, Eva. *Making a Beeline: My Journeys in Sixty Countries, 1949–2000.* Cardiff, UK: International Bee Research Association, 2003.

Crane, Eva. "Obituaries: Dr. Anna Maurizio, an Appreciation from IBRA." *Bee World* 75, no. 2 (1994): 98.

Crane, Eva. "Overview of a Career in Beekeeping." Interview, Woodside House, London, August 24, 2005.

Crane, Eva. *The Rock Art of Honey Hunters*. Cardiff, UK: International Bee Research Association, 2001.

Crane, Eva. *The World History of Beekeeping and Honey Hunting*. London: Routledge, 1999.

Crane, Eva. "The Worshipful Company of Wax Chandlers." *Bee World* 42, no. 3 (1961): 63–71.

Crane, Eva, and A. J. Graham. "Bee Hives of the Ancient World. 1." *Bee World* 66, no. 1 (1985): 23–41.

Crockett, Mrs. L. M. "'Beekeeping for Women.'—Read at Maine Convention." *American Bee Journal* 21, no. 30 (1885): 473–74.

Crozier, Ching. "Overview of Life as a Researcher." E-mail, July 3, 2007.

Crozier, Ross. "Overview of Life as a Researcher." E-mail, July 5, 2007.

Dadant, Charles. "Down with the Importation of Bees." *American Bee Journal* 11, no. 3 (1875): 54–55.

Dadant, Charles. "Italian Bees." *American Bee Journal* 11, no. 7 (1875): 149.

Dadant, Charles. "Our Foreign Bee Notes." *American Bee Journal* 11, no. 8 (1875): 177–78.

Davis, Merlene. "Equal Pay Should Be a Reality, Not a Dream." *Lexington Herald-Leader*, April 22, 2010, A3, A7.

de Guzman, Lilia. "Growing Up in the Philippines." E-mail, May 13, 2010.

diGrandi-Hoffman, Gloria. "A Career as a Researcher." Interview, January 15, 2011.

diGrandi-Hoffman, Gloria, Diana Sammataro, and Ruben Alarcon. "The Importance of Microbes in Nutrition and Health of Honey Bee Colonies: Factors Affecting the Microbial Community in Honey Bees." *American Bee Journal* 149, no. 7 (2009): 667–69.

Dodd, Victor. *Beemasters of the Past*. Mytholmroyd, UK: Northern Bee Books, 1983.

Dolbeare, B., ed. *A Narrative of the Captivity and Suffering of Dolly Webster among the Comanche Indians in Texas*. Clarksburg, VA: M'Granahgan and M'Carty, 1843. Reprint, New Haven, CT: Yale University Press, 1986.

Douglas, Ann. *The Feminization of American Culture*. New York: Doubleday, 1988.

Douglas, J. C. *A Handbook of Beekeeping for India*. Calcutta: Superintendent of Government Printing, 1884.

Dumas, Clark Griffith. "Apiculture in Early Texas." MS thesis, Southern Methodist University, TX, 1952.

Dummelow, J. *The Wax Chandlers of London: A Short History of the Worshipful Company of Wax Chandlers.* London: Phillimore, 1973.

Dunbar, W. "Memoir of Huber." *American Bee Journal* 2, no. 4 (1866): 61–63.

Dunham, Frances. "Langstroth." *American Bee Journal* 15, no. 7 (1879): 321.

Earle, Alice Morse, ed. *Diary of Anna Green Winslow: A Boston School Girl of 1771.* Boston: Houghton, Mifflin, 1894.

Ede, Piers Moore. *Honey and Dust: Travels in Search of Sweetness.* London: Bloomsbury, 2006.

E.J.P. [pseud.]. "Beekeeping for Women." In *Beekeeping for Women,* ed. A. I. Root, 1–3. Medina, OH: A. I. Root, 1906.

Ellis, Marion. "Early History of the Nebraska Beekeepers Association: Mrs. J.N. Heater." *Bee Tidings* (December 2001). http://entomology.unl.edu/beekpg/tidings/btid2001/btddec01.htm.

Embrey, Mike, and Bill Lord. "A Turkman Honey Tale." *Bees for Development Journal* 62, no. 9 (2003): 4–5.

Engel, M. S., Ismael Hinojosa-Diaz, and Alexandr Rasnitsyn. "A Honey Bee from the Miocene of Nevada and the Biogeography of *Apis* (Hymenoptera: Apidae: Apini)." Paper presented at the Proceedings of the California Academy of Sciences, San Francisco, May 7, 2009.

Fauls, H. "Voices from the Hives." *American Bee Journal* 11, no. 1 (1875): 21.

F.B. *The Office of the Good Housewife.* London: T. Ratcliffe and Richard Mills, 1672.

Ferrara, Cos. *A Love Affair with Nature: The Story of Anna Botsford Comstock, Pioneering Naturalist, Artist, Writer, and Teacher.* Basking Ridge, NJ: Girls Explore, 2004.

Fife, Austin. "The Concept of the Sacredness of Bees, Honey and Wax in Christian Popular Tradition." PhD dissertation, University of California, 1939.

Fisher, K. *In the Beginning, There Was Land: A History of Washington County, Iowa.* Washington, IA: Washington County Historical Society, 1978.

Fletcher, T. Bainbridge. *Beekeeping.* Edited by C. C. Ghoush. Calcutta: Superintendent Government Printing, 1915.

Flottum, Kim. *The Backyard Beekeeper: An Absolute Beginner's Guide to Keeping Bees in Your Yard and Garden.* Backyard Series. Gloucester, MA: Rockport, 2005.

Flottum, Kim. "Jennifer Berry . . . Queen Producer." *Bee Culture* 138, no. 2 (2010): 42–45.

Flottum, Kim. "Marla Spivak Wins MacArthur Fellowship." *Bee Culture* 138, no. 11 (2010): 37.

Flottum, Kim. "Winding Down at the White House." *Bee Culture* 147, no. 11 (2009): 42.

"Forgeries by a Woman: An Agricultural Writer in Iowa a Humble Imitator of E.D. Winslow." *New York Times,* February 5, 1876.

Forman, Elaine Crane, ed. *The Diary of Elizabeth Drinker: The Life Cycle of an Eighteenth-Century Woman.* Abridged ed. Philadelphia: University of Pennsylvania Press, 2010.

Fraser, M. *History of Beekeeping in Britain.* London: International Bee Research Association, 1958.

Frazier, Maryann. "Experiences as Bee Inspector in US and Africa." Interview, Heartland Apiculture Society conference, Southern Illinois University, Edwardsville, July 2005.

Frazier, Maryann. "Update on Colony Collapse Disorder in Honey Bee Colonies in the United States." Washington, DC: Committee on Agriculture, Subcommittee on Horticulture and Organic Agriculture, 2008.

Friedman, Thomas. "Letter from Istanbul, Part Two." *New York Times,* June 19, 2010, WK8.

Galton, Dorothy. "Beeswax as an Import in Mediaeval England." *Bee World* 52, no. 2 (1971): 68–74.

Galton, Dorothy. *Survey of a Thousand Years of Beekeeping in Russia.* London: International Bee Research Association, 1971.

Geithmann, Harriet. "The Story of Lillian Hill, Beekeeper of the Elbe." *American Bee Journal* 66, no. 4 (1936): 167–68.

Gentry, Nancy. "National Honey Standard." Interview, American Beekeeping Federation conference, Orlando, FL, January 15, 2010, with e-mail and telephone follow-up.

Gerald, Carolyn. "The Black Writer and His Role." In *The Black Aesthetic,* ed. Addison Gayle, 370–78. Garden City, NJ: Doubleday, 1971.

Glatz, Roberta. "My Life as a Beekeeper." Interview, Eastern Apiculture Society conference, Newark, DE, August 10, 2007.

Glushkov, N. M. "Facts about Beekeeping in the U.S.S.R." *Bee World* 40, no. 7 (1959): 169–72.

Glushkov, N. M. "Facts about Beekeeping in the U.S.S.R. (Continued)." *Bee World* 40, no. 8 (1959): 201–4.

Graham, Joe. "Burr Comb: Women in Beekeeping." *American Bee Journal* 119, no. 6 (1979): 460–63.

Graham, Joe. "Florida Adopts Honey Identity Standard; Other States Urged to Do Likewise." *American Bee Journal* 148, no. 4 (2008): 287.

Graham, Joe, ed. "World Honey Market." *American Bee Journal* 148, no. 9 (2008): 779–81.

Greenwood, Helen. "Bees on Their Knees." *Sydney Morning Herald,* June 24, 2008.

[Griffith, Mary]. "On Bees." *North American Review* 27, no. 61 (1828): 338–59.

Grozinger, Christina. "Research with Queen Pheromones." Telephone interview, April 29, 2010.

Guest, Edith. "Ballyvourney and Its Sheela-Na-Gig." *Folklore* 48, no. 4 (1937): 374–84.

Guy, R. D. "A Commercial Beekeeper's Approach to the Use of Primitive Hives." *Bee World* 52, no. 1 (1971): 18–24.

Hale, Sara Josepha Buell. "Literary Notices." *Ladies Magazine* 2 (1829): 199–200.

Hanaford, Phebe. *Daughters of America; or, Women of the Century.* 4th ed. Augusta, ME: True and Company, 1883.

Hans-Ulrich, Thomas. "Retrospect: Christian Konrad Sprengel and His Book *The Discovered Secret of Nature.*" *Bee World* 84, no. 1 (2003): 44–50.

Harrison, Mrs. Lucinda. "Excelsior—a Lady's Experience." *American Bee Journal* 15, no. 7 (1879): 295.

Harrison, Mrs. Lucinda. "The International Congress." *American Bee Journal* 21, no. 3 (1885): 35.

Harrison, Mrs. Lucinda. "Lady Presidents." *American Bee Journal* 23, no. 4 (1887): 58.

Harrison, Mrs. Lucinda. "Maddened Bees-Robbing." *American Bee Journal* 15, no. 6 (1879): 245.

Harrison, Mrs. Lucinda. "Watch the Swarms." *American Bee Journal* 15, no. 9 (1879): 407.

Henson, Pamela. "The Comstock Research School in Evolutionary Entomology." *Osiris,* vol. 8, 159–77. Special Topics: Research Schools Historical Reappraisals. Chicago: University of Chicago Press, 1993.

Hoffman, Marc. "Eureka! On the Streets of Philadelphia." *Bee Culture* 138, no. 4 (2010): 29–33.

Holcombe, Elaine. "The Importance of the Honey Queen Program." Interview, Shelbyville, TN, April 1, 2007.

"Honey Bees to Bring Money for Women." *Hindu,* October 14, 2007. http://www.hindu.com/2007/10/14/stories/2007101455700900.htm.

Horn, Tammy. *Bees in America: How the Honey Bee Shaped a Nation.* Lexington: University Press of Kentucky, 2005.

Hughes, Whitney. "How Sweet It Is." Airborne website, April 9, 2010. http://www.cjtf101.com/regional-command-east-news-mainmenu-401/2639-story-by-us-army-staff-sgt-whitney-hughes-task-force-wolverine-public-affairs.html.

Instruction in Bee-Keeping for the Use of Irish Beekeepers. London: His Majesty's Stationery Office, 1912.

Ioyrish, Naum. *Bees and People.* Edited by H. C. Creighton. Moscow: MIR, 1974.

Irving, Washington. *Diedrich Knickerbocker's History of New York: From the Beginning of the World to the End of the Dutch Dynasty*. New York: G. P. Putnam's Sons, 1849.

Islam, Md. Nurul. "Beekeeping in Bangladesh." In *Asian Bees and Beekeeping: Progress of Research and Development*, ed. M. Matsuka, 271–72. Enfield, NH: Science Publishers, 1998.

Jackson, Jack. *Comanche Moon*. New York: Reed, 1979. Reprint, 2003.

Jasim, Khalel. "Beekeeping in Iraq." *Modern Beekeeping* 36, no. 12 (1952): 405–7, 414.

Jay, Ricky. *Extraordinary Exhibitions: The Wonderful Remains of an Enormous Head, the Whimisphusicon and Death to the Savage Unitarians*. New York: Quantuck Lane, 2005.

Jefferson, Thomas. *Notes on the State of Virginia*. Edited by William Peden. Chapel Hill: University of North Carolina Press for the Institute of Early American History and Culture, Williamsburg, VA, 1955. http://etext.virginia.edu/toc/modeng/public/JefVirg.html.

Johansson, T. S. K., and M. P. Johansson. "A Letter from François Huber." *Bee World* 51, no. 4 (1970): 170–72.

Jones, Richard. "Dr. Eva Crane." *Eastern Apiculture Society Journal* 35, no. 4 (2007): 11.

Jones, Richard. "Further Notes on Eva Crane's Accomplishments." E-mail, April 20, 2010.

Jutte, Robert. *Poverty and Deviance in Early Modern Europe*. Cambridge: Cambridge University Press, 1994.

Kerr, R. "My Life." Unpublished manuscript, Ramona Historical Association, Ramona, CA, 1935.

Kessler, Cristina. *The Best Beekeeper in Lalibela*. New York: Holiday House, 2006.

Keyes, Mrs. A. S. "Letters in Our Box." *American Bee Journal* 15, no. 7 (1879): 302.

Khalifman, Iosif Aronovich. *Bees: A Book on the Biology of the Bee Colony*. Moscow: Foreign Language Publishing House, 1951.

Kigatiira, K. I. "Bees and Beekeeping in Kenya." *Bee World* 65, no. 2 (1984): 74–80.

Klarer, Mario. "Women and Arcadia: The Impact of Ancient Utopian Thought on the Early Image of America." *Journal of American Studies* 27, no. 1 (1993): 1–17.

Knight, Juliet. "Cath Keay: A New Destiny Is Prepared." *Hi Arts,* August 21, 2006. http://www.hi-arts.co.uk/Default.aspx.LocID-hianewlz8.RefLocID hiacg5005001.Lang-EN.htm.

Knowles, David, ed. *Medieval Religious Houses: England and Wales*. London: Longmans, Green, 1953.

Kritsky, Gene, ed. *The Quest for the Perfect Hive: A History of Innovation in Bee Culture.* Oxford: Oxford University Press, 2010.

Kupperman, Karen Ordahl. "Beehive as Model for Colonial Design." In *America in European Consciousness 1493–1750,* ed. Karen Ordahl Kupperman, 272–94. Chapel Hill: University of North Carolina Press, 1995.

Kupperman, Karen Ordahl. "Death and Apathy in Early Jamestown." *Journal of American History* 66, no. 1 (1979): 22–40.

Kupperman, Karen Ordahl. *Facing Off: Encounters of English and Indian Cultures in America 1640–1700.* Chapel Hill: University of North Carolina Press, 2000.

Lachman, Richard. *From Manor to Market: Structural Change in England, 1563 to 1640.* Madison: University of Wisconsin Press, 1987.

Laidlaw, Harry, Jr., and Robert Page Jr. *Queen Rearing and Bee Breeding.* Edited by Larry Conner. Cheshire, CT: Wicwas, 1997.

Landim, Patrícia. "EU Honey Legislation: EU Suspends Import of Brazilian Honey." *Bees for Development Journal* 79, no. 6 (2006): 6.

Langstroth, Lorenzo. "The Hive and the Honey Bee." 3rd ed. New York: C. M. Saxton, 1862. Reprinted in *The Hive and the Honey Bee Revisited,* annotated by Roger Hoopingarner. Holt, MI: Bee Plus Books, 2006.

Langstroth, Lorenzo. "Langstroth's Reminiscences." Unpublished papers. E. F. Phillips Beekeeping Collection, Cornell University Mann Library, Ithaca, NY.

Langstroth, Lorenzo. "Langstroth's Reminiscences: Getting the Movable Frame Introduced, Invention of the Honey Extractor, etc." *Bee Culture* 21, no. 6 (1893): 206–7.

Langstroth, Lorenzo. "Langstroth's Reminiscences: How He Became Interested in Bees." *Bee Culture* 21, no. 3 (1893): 80–81.

Langstroth, Lorenzo. "Letter to Charles Dadant." Oxford, OH, July 7, 1883. Langstroth-Dadant Correspondence, 1881–1888. E. F. Phillips Beekeeping Collection, Cornell University Mann Library, Ithaca, NY.

Lehman, Kim. "Kids and Bees Programs." E-mail, April 19, 2010.

Levi, Ed. "Travels in India and Nepal." Interview, Franklin, TN, October 19, 2007.

Linswick, Cyula. "Nellie's Experiment." *American Bee Journal* 11, no. 12 (1875): 274–75.

Linswick, Cyula. "Shall Women Keep Bees.—Read before the Michigan Convention." *American Bee Journal* 13, no. 5 (1877): 170–72.

Marchese, Marina. *Honeybee: Lessons from an Accidental Beekeeper.* New York: Black Dog and Levanthal, 2009.

Marchese, Marina. "Montalcino: The City of Honey." E-mail, June 1, 2010.

"Margaret James Murray Washington." Biography. Alabama Women's Hall of Fame website. http://www.awhf.org/washington.html.

Margulis, Lynn. *The Symbiotic Planet.* New York: Phoenix, 1998. Reprint, 2001.

Marshall, Elizabeth, and Samuel Marshall. "Letter." 1876. San Diego Museum of History, San Diego, CA.

Mathie, George. "Woman Beekeeper 'Sees It Through.'" *Farmer's Weekly,* January 20, 1932, 27–28.

McAdam, Betty. "Being a Beekeeper on Kangaroo Island." E-mail, August 2, 2008.

McConnell, Miss Belle. "Personal Experience of a Teacher and Beekeeper." In *5th Annual Report of the State Bee Inspector and the Report of the Convention of the Iowa State Beekeepers' Association, Dec. 5 & 6, 1916,* ed. Frank Pellett, 49–50. Des Moines: Iowa State Beekeeper's Association, 1917.

McDonough, Margo. "Building a Better Bee." *University of Delaware Messenger* 16, no. 1 (2008). http://www.udel.edu/PR/Messenger/07/01/building.html.

McGaughey, Steve. "Grozinger Found Her Inspiration at Illinois." *Synergy* (Summer 2007). Beckman Institute for Advanced Science and Technology website. http://www.beckman.illinois.edu/synergy/Summer 2007/grozinger.

McGregor, Liann. "Forty Years as a Beekeeper in South Africa." Interview, Durbin, South Africa, November 3, 2006.

McLoughlin, Robert. "Saint Gobnait." E-mail, August 28, 2005.

McNeil, M. E. A. "Bee Time: Ancient, Reversible, and Counting on a New Era." *American Bee Journal* 148, no. 6 (2008): 539–43.

McPherson, Misty. "Argentina Couple Learning about Beekeeping in the United States." *Claxton Enterprise,* August 6, 1992.

Mickels-Kokwe, Guni. *Small-Scale Woodland-Based Enterprises with Outstanding Potential: The Case of Honey in Zambia.* Jakarta: Center for International Forestry Research, 2006.

Mickels-Kokwe, Guni. "Zambia Beekeeping." Interview, Apimondia International Bee Congress, Melbourne, Australia, September 12, 2007.

Miller, Rita Skousen. *Sweet Journey: Biography of Nephi E. Miller.* Colton, CA: Miller Family Trust, 1994.

Mithen, Steven. *After the Ice: A Global Human History 20,000–5000 B.C.* Cambridge, MA: Harvard University Press, 2003.

Monaghan, Patricia. "Austēja." Edited by Normandi Ellis. E-mail, May 1, 2007.

Morandin, Lora. "Wild Bees and Agroecosystems." Student Apiculture Award, winner's address, presented at the Eastern Apiculture Society, Kent, OH, August 3, 2005.

More, Daphne. *The Bee Book: The History and Natural History of the Honey Bee.* New York: Universe Books, 1976.

Morgenthaler, O. "In Memory of Miss Annie D. Betts." *Bee World* 42, no. 12 (1961): 307–13.

Mutinelli, Franco, Cecilia Costa, Marco Lodesani, Alessandra Baggio, Piotr Medrzycki, Giovanni Formato, and Claudio Porrini. "Honey Bee Colony Losses Italy." *Journal of Apicultural Research: Colony Losses* 49, no. 1 (2010): 199–200.

Nash, Charles Elventon, ed. "Mrs. Ballard's Diary, 1785–1812." In *The History of Augusta: First Settlements and Early Days as a Town,* 229–463. Augusta, ME: Charles E. Nash and Son, 1904.

Neumann, Peter, and Norman Carreck. "Honey Bee Colony Losses." *International Bee Research Association* 49, no. 1 (2010): 1–6.

Newman, Thomas. "Apicultural News Items—Historical Item about Madame De Padua." *American Bee Journal* 21, no. 23 (1885): 355.

Newman, Thomas. "Canadian Apiarists in London." *American Bee Journal* 22, no. 46 (1886): 728–29.

Newman, Thomas. "Ellen Tupper—Obituary." *American Bee Journal* 24, no. 12 (1888): 179.

Newman, Thomas. "Lady Beekeepers: Anonymous Letter to Which Newman Responds." *American Bee Journal* 15, no. 12 (1879): 523.

Newman, Thomas. "Lizzie Cotton." *American Bee Journal* 24, no. 10 (1888): 147.

Newman, Thomas. "Mrs. J.N. Heater." *American Bee Journal* 37, no. 14 (1897): 209.

Newman, Thomas. "Mrs. Tupper's Troubles." *American Bee Journal* 12, no. 10 (1876): 252.

Newman, Thomas. "Notes and Queries." *American Bee Journal* 13, no. 4 (1877): 120.

Newman, Thomas. "Notes and Queries—Ellen Tupper." *American Bee Journal* 12, no. 4 (1876): 85.

Newman, Thomas. "Notes and Queries—Mrs. F.A. Dunham's Wax Foundation." *American Bee Journal* 15, no. 1 (1879): 41.

Newman, Thomas. "Notes and Queries—Spaid Honey House." *American Bee Journal* 12, no. 3 (1876): 56.

Newman, Thomas. "The Queen Bee's Temptation." *American Bee Journal* 12, no. 3 (1876): 53–54.

Noel, Mrs. Clara T. "My Bees." Paper presented at the Iowa Beekeepers Convention, Des Moines, 1917.

Oberste, W. *Texas Irish Empresarios and Their Colonies.* Austin, TX: von Boeckman-Jones, 1953.

O'Brien, Tom. "Ontario's Tech Transfer Team: Meet the Bee Girls!" *Bee Culture* 138, no. 2 (2010): 47–49.

Oldroyd, Benjamin, and Siriwat Wongsiri. *Asian Honey Bees: Biology, Conservation, and Human Interactions.* Cambridge, MA: Harvard University Press, 2006.

Ostiguy, Nancy. "A Career in Bee Research." E-mail, May 10, 2010.

Ostiguy, Nancy. "Managed Pollinator CAP: Sustainable Beekeeping." *American Bee Journal* 150, no. 2 (2010): 149–52.

Ostiguy, Nancy. "New Integrated Pest Management Techniques in Beekeeping." Presentation at Heartland Apiculture Society conference, Southern Illinois University, Edwardsville, July 8, 2005.

Padilla, F., F. Puerta, J. M. Flores, and M. Bustos. "Bees, Apiculture and the New World." *Archivos de zootecnia* 41, no. 154 (1992): 563–67.

Pantziara, Nicoletta. "From Ancient to Modern: Greek Women's Struggle for Equality." *Social Education* 67, no. 1 (2003): 28–31.

Park-Burris, Jackie. "Overview of Park-Burris Family History." Interview, May 15, 2010.

Pasco, Richard. "Transshipped Honey Hurting U.S. Beekeepers." *Catch the Buzz,* February 2, 2010. http://home.ezezine.com/1636/1636-2010.02.02.12.53 .archive.html.

Patterson, Daniel. *The Shaker Spiritual.* Princeton, NJ: Princeton University Press, 1979.

Peiss, Kathy. *Hope in a Jar: The Making of America's Beauty Culture.* New York: Henry Holt, 1998.

Pellett, Frank. "Honey Production in the Sage District: Notes on the Methods of a Well-Known Beekeeper Who Produces Honey on a Large Scale." *American Bee Journal* 59, no. 9 (1919): 295–96.

Pellett, Frank. "How Women Win: Three School Teachers Who Have Become Very Successful Beekeepers." *American Bee Journal* 57, no. 11 (1917): 372–73.

Pellett, Frank. "Mrs. C.C. Miller and Miss Wilson Die." *American Bee Journal* 73, no. 5 (1933): 169.

Phelps, Dr. W. G. "Women as Beekeepers." *American Bee Journal* 21, no. 20 (1885): 315.

Phillips, E. F. "Beekeeping in the Post-war Era." In *Report of the Iowa State Apiarist for the Year Ending 1944,* 49–54. Des Moines: Iowa State Govt., 1945.

Phillips, E. F. "War Time Regulations Affecting the Beekeeping Business." In *Iowa State Apiarist Report for the Year Ending 1941,* 58–62. Des Moines: Iowa State Govt., 1942.

Phipps, Ron. "International Honey Market Report." *American Bee Journal* 148, no. 5 (2008): 399–401.

Phipps, Ron. "International Honey Market Report." *American Bee Journal* 147, no. 12 (2007): 1016.

Prescott, James Arthur. "The Russian Free Economic Society: Foundation Years." *Agricultural History* 51, no. 3: 503–12.

Prete, Frederick. "Can Females Rule the Hive? The Controversy over Honey Bee Gender Roles in British Beekeeping Texts of the Sixteenth–Eighteenth Centuries." *Journal of the History of Biology* 24, no. 1 (1991): 113–44.

Pundyk, Grace. "Continuing the Honey Spinner." E-mail, May 10, 2010.

Pundyk, Grace. "Further Recollections on Turkey." E-mail, May 11, 2010.

Pundyk, Grace. *The Honey Spinner: In Pursuit of Liquid Gold and Vanishing Bees.* New York: St. Martins, 2010.

Ransome, Hilda. *The Sacred Bee in Ancient Times and Folklore.* London: George Allen and Unwin, 1937.

Reid, T. R. *Confucius Lives Next Door: What Living in the East Teaches Us about Living in the West.* New York: Random House, 1999.

Roberts, Mildred. "Ann Gudgel." In *Kentucky Slave Narratives: A Folk History of Slavery in Kentucky with Interviews with Former Slaves,* prepared by Federal Writers' Project, 1936–1938, 28. Bedford, MA: Applewood Books, 2006.

Rocheblave, Jacqueline. "A French Woman's Experience as a Beekeeper." Interview, Apimondia International Bee Congress, Melbourne, Australia, September 13, 2007.

Root, A. I. "Headlines—Ellen Tupper." *Bee Culture* 4, no. 3 (1876): 54.

Root, A. I. "Honor Roll—Ellen Tupper." *Bee Culture* 3, no. 4 (1875): 5.

Root, A. I. "Humbugs and Swindlers—Letter Written by Will R. King." *Bee Culture* 1, no. 12 (1873): 93.

Root, A. I., ed. "Humbugs and Swindlers—Lizzie Cotton." *Bee Culture* 2, no. 4 (1874): 41.

Root, A. I. "Notes and Queries—A Kind Word for Mrs. Cotton." *Bee Culture* 14, no. 14: 588.

Root, A. I. "Notes and Queries—Letter from William Ellsworth." *Bee Culture* 7, no. 5 (1876): 117.

Root, A. I. "Roll Call: Miss Annie C. Mann." *Bee Culture* 3 no. 4 (1875): 6.

Runte, Roseann. "From La Fontaine to Porchat: The Bee in the French Fable." *Studies in Eighteenth-Century Culture* 18 (1988): 79–89.

Sammataro, Diana. "Family Beekeeping History." Interview, Dublin, Ireland, August 26, 2005.

Sammataro, Diana. "Research Past and Present." Telephone interview, March 20, 2010.

Sammataro, Diana, and Alphonse Avitabile. *The Beekeeper's Handbook.* 4th ed. Ithaca, NY: Cornell University Press, 2006.

Sanford, Malcolm. "The Africanized Honey Bee: A Biological Revolution with Human Cultural Implications, Part V—Conclusion." *American Bee Journal* 146, no. 7 (2006): 597–99.

Sanford, Malcolm. "The Florida Standard of Identity for Honey." *Bee Culture* 136, no. 12 (2008): 21.

Sanford, Malcolm. "The Fourth Individual in a Honey Bee Colony." *American Bee Journal* 134, no. 9 (2006): 17–19.

Sanford, Malcolm. "New Generation Honey Cooperatives: Has Their Time Come?" *Bee Culture* 146, no. 4 (2006): 17.

Satrapi, Marjane. *Persepolis 2.* New York: Pantheon, 2004.

Savery, Annie. "Experiences of a Beekeeper." Paper presented at the Transactions of the North American Beekeepers' Society, Cleveland, OH, 1871.

Seaton, Beverly. "Mary Griffith." In *American Women Writers: A Critical Reference,* vol. 2, ed. Taryn Benbow-Pfalzgraf, 148. Detroit: St. James, 2000.

Seaver, James. *A Narrative of the Life of Mrs. Mary Jemison.* Edited by June Namias. Minneapolis, MN: Filiquarian, 2007.

Seeley, T., and J. Tautz. "Worker Piping in Honey Bee Swarms and Its Role in Preparing for Liftoff." *Journal of Comparative Physiology A: Sensory, Neural, and Behavioral Physiology* 187, no. 8 (2001): 667–76.

Sehgal, Sitaram. *Hindu Marriage and Its Immortal Traditions.* New Delhi: Navyug, 1969.

Sharma, Kunal. "Women and Bees." E-mail, August 2, 2008.

Sharma, Kunal, and Snehlata Nath. *Honey Trails in the Blue Mountains.* Kotagiri, India: Keystone Foundation, 2007.

Shiva, Vandana. *Stolen Harvest: The Hijacking of the Global Food Supply.* Cambridge, MA: South End, 2000.

Showler, Karl. "In the Apiary: Having Fun with Bees (Part 17)." *Bee Craft* 87, no. 10 (2005): 30–31.

Showler, Karl. "Some Bee Books by Women Writers: Part Two 1905–1953." *Bee Craft* 85, no. 10 (2003): 22–23.

Showler, Karl. "Some Bee Books by Women Writers: Part Three 1965–1988." *Bee Craft* 85, no. 11 (2003): 23.

Simões, Zilá Luz Paulino. "A Career in Research." E-mail, September 24, 2008.

Sipe, Lawrence. "How Picture Books Work: A Semiotically Framed Theory of Text-Picture Relationships." *Children's Literature in Education* 29, no. 2 (1998): 97–108.

Slackman, Michael. "Stifled, Egypt's Young Turns to Islamic Fervor." *New York Times,* February 17, 2008, A1.

Sloan, Scott. "Afghan Businesswoman Meets U.S. Style." *Lexington Herald-Leader,* May 8, 2010, B9.

Soares, Christine. "Building a Future on Science." *Scientific American* 298, no. 2 (2008): 80–85.

Soza, Samuel. "The Buzz in Iraq." *Bee Culture* 138, no. 5 (2010): 30.

Spivak, Marla. "The Coordinated Agricultural Project (CAP) Grant Project: Honey Bee Medical Records." *Bee Culture* 138, no. 3 (2010): 41–44.

Spivak, Marla. "Minnesota Hygienic Bees." Interview, June 15, 2004.

Spivak, Marla. "Overview of Costa Rica Experience." E-mail, April 16, 2010.

Spivak, Marla. "Overview of South America Experience." E-mail, May 24, 2010.

Spivak, Marla. "Social Safety Networks for Families: A Global Perspective." E-mail, April 17, 2010.

Stewart, Cal. "Uncle Josh and the Honey Bees." Recording. Edison Records, 1919.

Stoll, Steven. *The Fruits of Natural Advantage: Making the Industrial Countryside in California*. Berkeley: University of California Press, 1998.

Stover, Mrs. B. "Visiting California's Apiaries." *American Bee Journal* 21, no. 15 (1885): 219.

Stowe, Mrs. N. L. "Some Kindly Hints to Young Women." *Bee Culture* 23, no. 1 (1895): 20.

Strachan-Severson, Valeri. "Overview about Strachan Queens." Interview, May 6, 2010.

Strong, Anna Louise. *The Stalin Era*. New York: Mainstream, 1956.

Sweet, C. L. "Hibernation and Pollen in 1764." *American Bee Journal* 21, no. 4 (1885): 58.

Tam, Janet. "The Bee Girls Tech Transfer Team." E-mail, May 12, 2005.

Tarwater, Stephanie. "Inspecting Bees in Florida." Interview, July 8, 2010.

Tarwater, Stephanie. "Migratory Beekeeping in TN." Interview, Heartland Apiculture Society conference, Southern Illinois University, Edwardsville, July 8, 2005.

Tennant, Kylie. *The Honey Flow*. London: Macmillan, 1956.

Thacher, J. A. *A Practical Treatise on the Management of Bees*. Boston: Marsh and Capen, 1829.

Traill, Catherine Parr Strickland. *Backwoods of Canada: Being Letters from the Wife of an Emigrant Officer*. London: Nattali and Bond, 1838.

Traynor, Kristin. "Bee Breeding around the World." *American Bee Journal* 148, no. 2 (2008): 135–39.

True, Jere, and Victoria Tupper Kirby. *Allen Tupper True: An American Artist*. San Francisco, CA: Canyon Leap, 2009.

Tudge, Colin. *The Time before History: 5 Million Years of Human Impact*. London: Scribner, 1996.

Tupper, Ellen. "Answer to Mr. Dadant." *American Bee Journal* 11, no. 3 (1875): 51–52.

Tupper, Ellen. "Beekeeping." In *Annual Report,* Board of Directors of the Iowa State Agricultural Society, 143–53. Des Moines, IA: Department of Agriculture, 1864.

Tupper, Ellen. "Journal" (1880). Smithsonian Arts Archives, Washington, DC.

Ulrich, Laurel Thatcher. *Good Wives: Image and Reality in the Lives of Women in Northern New England, 1650–1750.* New York: Vintage, 1991.

Verettoni, Sonia. "Queen Breeding in Argentina." E-mail, September 22, 2008.

Vogt, Bill. "Anna Botsford Comstock." *National Wildlife* 26, no. 6 (1988): 50.

Wakefield, Peter, and Peter Cook. "Sting in the Tale." Letter exchange between Peter Wakefield and Peter Cook about Olive Brittan. *Times Online,* 2005, http://www.timesonline.co.uk/tol/comment/letters/article510724.ece.

Waldstreicher, D. *Runaway America: Benjamin Franklin, Slavery, and the American Revolution.* New York: Hill and Wang, 2004.

Walker, Penny. "Bee Boles and Past Beekeeping in Scotland." *Review of Scottish Culture* 4 (1988): 105–17.

Walker, Penny. "Overview of a Career with International Bee Research Association." Interview, Gerrard's Cross, London, August 23, 2005.

Warder, Joseph. *The True Amazons; or, The Monarchy of Bees.* 6th ed. London: Pemberton, 1716.

Washington, Booker T., ed. *Tuskegee and Its People: Their Ideals and Achievements.* New York: D. Appleton, 1905.

Washington, Margaret Murray. "We Must Have a Cleaner Social Morality." Speech given at the Old Bethel AME Church, Charleston, SC, September 12, 1898. Published in *Charleston News and Courier,* September 13, 1898. Entire speech available online at http://www.blackpast.org.

Watson, James. *Bee-Keeping in Ireland: A History.* Dublin: Glendale for the Federation of the Irish Beekeeping Association, 1981.

Weaver, B. "A History of the Zach Weaver Family and Weaver Apiaries." In *The History of Grimes County: Land of Heritage and Progress,* ed. Grimes County Historical Commission, 568. Dallas, TX: Taylor, 1982.

Webb, Virginia. "World Honey Shows and Honey Standards." Telephone interview, May 10, 2010.

Weiss, Bill. "Federal Council of Australian Apiarist's Associations." *Australasian Beekeeper* 108, no. 12 (2007): 501.

Wells, A. E. *Men of the Honey Bee.* Adelaide: Rigby, 1971.

Whitney, W. J. "Bees in California." *American Bee Journal* 12, no. 1 (1876): 18–19.

Who's Counting: Marilyn Waring on Sex, Lies and Global Economics [film; 52-minute short version]. Canada: National Film Board of Canada, 1995.

Wien, H. Christian, Roberta J. Glatz, and Roger Morse. "Flowering, Pollination, and Fruit Set." In *Pumpkin Production Guide,* ed. Dale Ila Miles Riggs, 43–47. Ithaca, NY: Natural Resource, Agriculture, and Engineering Service, 2003.

Willard, F. E., and Mary Livermore. *American Women: Fifteen Hundred Biographies.* New York: Mast, Cromwell, and Kirkpatrick, 1899.

Williams, Michael. *Deforesting the Earth: From Prehistory to Global Crisis.* Chicago: University of Chicago Press, 2002.

Wilson, Emma. "Bee-Keeping for Women." *American Bee Journal* 43, no. 1 (1903): 8–9.

Wilson, Emma. "Our Bee-Keeping Sisters." *American Bee Journal* 54, no. 2 (1917): 95.

Winston, Mark. *The Biology of the Honey Bee.* Cambridge, MA: Harvard University Press, 1987.

Withington, Ann Fairfax. "Republican Bees: The Political Economy of the Beehive in Eighteenth-Century America." *Studies in Eighteenth-Century Culture* 18 (1988): 39–77.

Wollstonecraft, Mary. *A Vindication of the Rights of Woman.* Edited by Miriam Brody. London: Penguin, 2004. Orig. pub. 1792.

Woods, E. F. "Queen Piping." *Bee World* 37, no. 10 (1956): 185–95, 216–19.

Woodworth, Bonnie. "Migratory Beekeeping Challenges." Presentation at the American Beekeeping Federation conference, Louisville, KY, January 13, 2006.

Wordsworth, William. "Sonnet 33." In *Wordsworth: Poetical Works,* ed. Thomas Hutchinson, 206. Oxford: Oxford University Press, 1990.

Working, D. W. *Bees in Colorado.* National Beekeepers' Association, 1902.

York, George. "Miss Emma Wilson's Beekeeping Sisters Department." In *The Story of the American Bee Journal.* Chicago: Thomas Newman and Son, 1903.

Index

Page references followed by italicized *p* indicate photos; *b,* boxes.